To order additional copies of this book, contact:
Xlibris
1-888-795-4274
www.Xlibris.com
Orders@Xlibris.com

STATISTICS TOOLS

2nd Edition

Alice Gorguis
North Park University

Making Decisions

In Statistics, to investigate a statement, a **five-step procedure** called hypotheses-testing is applied to make the decision of: Rejecting or not rejecting the null hypotheses.

In our life, **five-step procedure** may be applied to make decisions:
- Pray before making decisions (Nehemiah 1:4)
- Decide to do things that honor God (Job 1:8)
- God's word helps us make decisions (Psalm 119:105)
- Get good advice before making decisions (Proverbs 18:15)
- Ask God for wisdom before making decisions (James 1:2-8)

Contents

P R E F A C E

This book is designed for students with minimal background in mathematics with a format that will work for a variety of students in an introductory statistics course. Problems are simplified by using step-by-step approach along with graphs to make it easier for students to understand. The use of technology EXCEL, TI-83, 84+, SPSS is explained next to each example to make it clear and easy to follow. Important formulas and notes are highlighted.

The primary objective of this book is to present an easy introduction to statistics to promote learning and understanding.

Statistics requires the use of many formulas, so those students who have not had intermediate algebra should complete at least one semester of college mathematics before beginning this course.

By Statistics Tools we mean the topics used in statistics that are presented in the book,

If the student understands the concept behind each topic, then he/she can use it as a tool to solve any problem in the real world.

Finally the book is designed to cover all the topics required by the school for one semester of Four credit hours.

In this second edition extra problems were added especially to chapters -8 and 9.and extra problems with explanation in other sections.

1 Introduction to Statistics

1.1 Descriptive Statistics
1.2 Levels of Data Measures
1.3 Collecting Data Samples
1.4 Errors in Statistics

Introduction

Statistics is the universal language of the sciences. With understanding data, and careful use of the statistical tools, scientists can be able to make accurate description of their findings of scientific research, and make decisions and estimations.

Most of the important questions in our lives involve incomplete information; statistics takes care of drawing a good, reliable conclusion from this information.

We are bombarded by statistical information everywhere: in newspapers, TV, internet.

Courses in statistics are required or recommended in Psychology, Sociology, Computer Science, Biology, Nursing, Business, Linguistics, Political Science, Education, Pre-Medicine, and Pre-Law. In short, you'll find statistics in any field that requires to present data in a meaningful way. Computing numbers meaningful in statistical analysis sometimes requires calculations, so in each section we present some information about using calculators (TI-84+), also using EXCEL, and SPSS to assist the student in performing the calculations necessary for the statistical analysis.

So statistics is used in almost all fields that conduct studies on subjects that require to: collect, organize, analyze data, and then draw conclusion from the data.

Statistics: Is the science of conducting studies that deals with collected data.

To study statistics we need to be able to speak the language. So we start first by defining basic terms that will be used throughout the course.

Data The set of values collected for the response variable from each element belonging to the sample, the value of the data can be symbolized by (x, y, z).

Population Is a collection of individuals, objects, or measurements that are of interest to statisticians, and is symbolized by (P).

Sample Group of objects selected from population, and is symbolized by (S).

Experiment A planned activity, whose outcome yields a set of data.

Statistics is divided into 2-main areas:
- Descriptive Statistics.
- Inferential Statistics.

1.1 Descriptive Statistics

Descriptive Statistics: Is the science of conducting studies to collect, organize, Summarize, analyze, and draw conclusion from data. In one word we can call it:

COSAD

C for collecting, O for organizing, S for summarizing, A for analyzing, and D for Draw conclusions.

Data: Two types of data are used qualitative data (non-numerical and measurable) and quantitative data or (numerical and countable), and the former one is divided into discrete data (with whole numbers), and continuous (with all points in between).

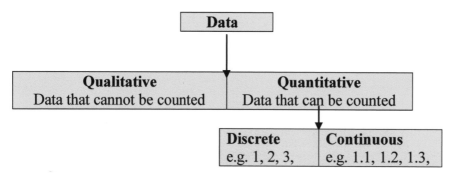

1.2 Levels of Data Measures

1.Nominal Level	Is the data that consists of names, labels, or data without order or ranking, such as: zip-code, gender, political affiliations, . . .etc.
2.Ordinal Level	Data that can be arranged in order, such as: grades (A, D,), ranking scale (poor, good, excellent).
3.Interval Level	Data that can be arranged in order, and do not have zero, such as: IQ, SAT score, temperature.
4. Ratio Level	Data that can be organized in order and has natural zero starting point

1.3 Collecting Data Samples

Statistical methods are driven by the data that we collect. Data can be collected in many Different ways as shown:

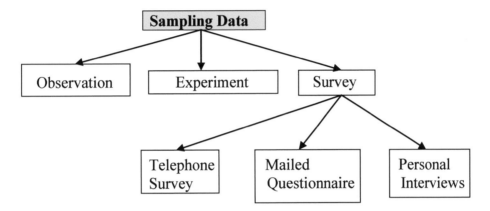

1.4 Errors in Statistics

Unlike Mathematics, in Statistics it is likely to be some errors in the result, and the type of errors that occur are:

<u>Sampling errors:</u> Occurs when the sample is selected with random method.

<u>Non sampling errors:</u> is the result of human errors.

<u>Standard errors:</u> Type-I error (probability of α) = Rejecting a true null hypotheses.

Type-II error (probability of β) = Accepting a false null hypothesis.

The two types of errors are inversely related, as α decreases β increase.

2 Organizing Data

2.1 Constructing Frequency Distribution Tables
2.2 Constructing Graphs
2.3 Practice Problems

Organizing Data

After collecting data, we have to organize it using two different methods:
1. Organizing data using tables called frequency distribution tables.
2. Organizing data using graphs.

No matter what type of data is used in the statistical studies, sample or population, it is important to organize it and present it to the public in a meaningful way such as tables, or graphs. In this chapter we will concentrate on frequency distribution for tables, and Histograms; Frequency-Polygons, and O-Gives for the graphs.

> **Florence Nightingale:** was a pioneer in using graphs to present data in a vivid form that people could understand. Her inventive graphs were a landmark in the growth of the new science of Statistics. She considered Statistics essential to understanding any social issue.

2.1 Constructing Frequency Distribution Tables

Two types of frequency distribution is presented in this section, one that deals with Qualitative data or data that can be measured, such as height (tall, medium, short), Gender (Boy, girl) and blood pressure (A, B, AB, O) ...etc. The other type is the one that deals with Quantitative data, or data that can be counted, and are divided into: discrete, and continuous.

A. Frequency Distribution for Non-Numerical (qualitative) data:
The qualitative data are the data that can be measured such as gender, blood Pressure...etc.

Example-1 In a hospital the date for a sample of blood type for 30-people shows:

Data: A O B A AB B O B O A O B O O AB
A O B B O B O AB A B B O AB AB A

Construct a frequency distribution for the given data.

Solution:

> **To construct a frequency distribution means to organize the date in a table form, using classes and frequencies, often time % is used too.**

Data	Frequency (f)	Percent = f/n x 100%
A	6	6/30 x 100% =20%
B	9	9/30 x 100% = 30%
O	10	10/30 x 100% ≈ 33%
AB	15	15/30 x 100% ≈ 17%
	Σ= n = 30	Σ = 100%

The construction shows that, more people are of type-O than any other type.
Note: in case of non-numerical data, the percentage is often times used.

B. Frequency Distribution for Numerical (quantitative) data:

As mentioned before, quantitative data are the data that can be counted, and are divided into: Discrete, and continuous. If the value of data, or range of the numbers, is small, then ungrouped frequency distribution is applied, but if the value of data, or the range of the numbers, is very large, then the grouped frequency distribution is applied.

1. Ungrouped frequency Distribution: Means each value of the data(x) in the distribution represents only one value, and data has a small range.

Example-2 A group of students from Stat-Class conducted the following experiment: Put 5-coins in a cup, shake it, and dump it out, count the number of heads up, and Record it. Repeat 20 times. The following data were formed:

Raw Data: 2 1 3 3 1 2 2 2 3 2
2 3 4 4 3 2 0 2 1 4

Construct the frequency distribution for the data.

> **Note: Raw date means unorganized data.**
> **Ungrouped data are the data with a small range.**

Solution: This data has a small range = highest vale (3) – lowest value (0) = 3

> As done in example-1 we organize the date in a table using classes and frequencies.

Data (x) # of Heads	Frequency (f)
0	1
1	3
2	8
3	5
4	3
	$\sum$ = n= 20

Symbols $\implies$ | Σ **means sum** | **n means sample size** |

2. Grouped frequency Distribution: When a large set of date has many different values in it rather than many repeated values as in the previous example, then the data are classified into a set of classes that will allow us to construct a frequency-distribution. As illustrated in the following example.

Constructing frequency distribution for a grouped data requires the following:

Class boundaries: are numbers used to separate the classes from each other so that there are no gaps in the frequency distribution. Classes must be mutually exclusive or non- overlapping class limits so that data cannot be placed into 2-classes.

Example-3   The following table shows how to find the class boundaries for different data with different percentage.

Data	Class (value)	(−) on L, (+) on R	Boundaries
Length	18 inch	$(18 - 0.5) - (18+0.5)$	$17.5 - 18.5$
Time	0.34 seconds	$(0.34 - 0.005) - (0.34 + 0.005)$	$0.335 - 0.345$
Mass	1.3	$(1.3 - 0.05) - (1.3+0.05)$	$1.25 - 1.35$

Example-4   For the following classes, find the class-boundaries, mid-points, and Class-width:

Class-limits
a. $32 - 45$
b. $11.4 - 13.6$
c. $2.19 - 5.75$

Solution:	Class-boundaries: a. ± 0.5; b. ± 0.05, c. ± 0.005. Mid-point = ½(sum of the two data)

Class-boundaries	Mid-points	Width
31.5 – 45.5	½(31.5+45.5) = 38.5	45.5 – 31.5 = 14
11.35 – 13.65	½(11.35+13.65) = 12.5	13.65 – 11.35 = 2.3
2.185 – 5.755	½(2.185+5.755) = 3.97	5.755 – 2.185 = 3.57

Number of classes: (C) although there is no rule in choosing number of classes contained in the frequency distribution, but it is important to have enough classes to present the collected data. As a rule we can choose number of classes between: 5 – 20.

Range: R = highest value – lowest value.

Class width: $W = \dfrac{\text{Range}}{\text{Number of classes}} = \dfrac{R}{C}$

The class-width W is found by subtracting the upper value of one class from the lower value of the next class, in a class-limits (in a vertical position), or subtracting the upper value of the class from the lower value of the same class, in a class-boundaries (in horizontal position).

Example-5	Find the width for the following class-limits, and class-boundaries.

Class-limits	Width
23 – 30	31 – 23 = 8
31 – 38	

Vertical position

Class-boundaries	Width
22.5 - 30.5	30.5 – 22.5 = 8

Horizontal position

Example-6	If the data is given in whole number, and the class width W = 4.9 then it should be rounded-up to the nearest whole number = 5. Even if the width W = 10.145 it should be rounded-up to → 11.

Example-7	In records of 50- students who completed 4-years of college in each state, data was listed as: (in % rounded up to whole digit)

Raw Data: 23 26 30 34 26 22 31 22 23 25 24 38 33
24 24 24 21 22 24 28 29 25 24 17 36 28
24 21 24 21 25 19 23 37 33 30 25 26 25
25 24 27 28 26 27 27 15 34 28 31

Construct the grouped frequency distribution for the data.

Solution:

To construct a frequency distribution for grouped date we organize
the date in a table using classes (limits, and boundaries), frequencies,
cumulative frequency, and mid-points.

First we gather the following information's:
Number of data = n = 50
Highest data value = HV = 38
Lowest data value = LV = 15
Range = HV – LV = 38 – 15 = 23
Number of classes = 5 (our choice between 5 to 20)
Width W = 23/5 = 4.6 round up to exact digit = 5 the width of each block.
Frequency = f = number of times each data appears in the set.
Mid-point of each class= X_m = (lower boundary + upper boundary) / 2
Cumulative-frequency (CF).

Now we will organize the information in a table form as follows:

Class-limits	Class-Boundaries	Frequency (f)	C-Frequency	Mid-point X_m
1. 15 – 19	14.5 – 19.5	3	0	17
2. 20 – 24	19.5 – 24.5	18	3	22
3. 25 – 29	24.5 – 29.5	18	21	27
4. 30 – 34	29.5 – 34.5	8	39	32
5. 35 – 39	34.5 – 39.5	3	47	37
		$\sum = n = 50$	50	

This table is called the frequency distribution table.

2.2 Constructing Graphs

After organizing data in tables called **"frequency distributions table",** now we can present
them to public (viewers) in a graphical form. The most common graphs are:
  a. Histogram with axes (x,y) represented by: (x=class-boundaries, y=frequency)=(CB,F)
  b. Frequency Polygon with axes (x,y) represented by: (x_m= class mid-points, y = frequency)
      = (MP, F)
  c. O-give with axes (x,y) represented by:(x = upper-class-boundaries, y = cumulative
      frequency) = (CB, CF)
  d. Pie-Graph.
To graph we will use TI-Calculators, EXCEL, and SPSS.

EXCEL

| Example-8 | Using data and information from example-7 above, construct a Histogram, Frequency polygon, and O-Give. |

a. Histogram

The above histogram graph was constructed using EXCEL in the following
Steps:
1. Enter all the 50-data in column-A of EXCEL sheet, one data in each cell.

2. Enter the upper boundaries in column-B.

3. Select Data Analysis from Data tool bar; if not in, use Ad-Ins to add it:
    To get ad-ins go to file→options→ad-ins.
    From Data-Analysis, select Histogram, and Click OK.

4. In the Histogram Box type A1:A50 in Input-Range box, and B1:B5 in Bin-Range box.
    Select Chart-Output, and click OK.

You will get a histogram with blocks separated by gaps, and the horizontal title is Bin.
1. To get rid of the gaps in between the blocks, place the mouse inside a block and R-CLICK, then change the gap to –zero.

2. To change the title from Bin to Class-Boundaries click the mouse anywhere in the graph and you get a chart for axis and titles.

b. frequency Polygon:

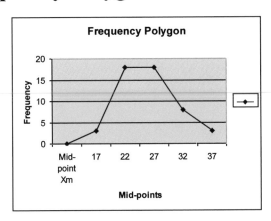

This graph was constructed using EXCEL in the following steps:
1. Highlight the two columns: Mid-Point and Frequency.
2. Click on Insert then Chart, then chose Line, then click NEXT.
3. Click NEXT to insert the titles and axes.

c. O-give:

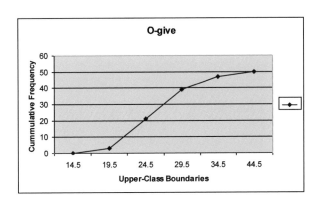

This graph was constructed using EXCEL in the following steps:
1. Insert the upper-class-boundaries including the first data in a column of cells, and cumulative frequency (starting with 0) in the next column of cells.
2. Highlight both columns. Click on Insert then Chart, then chose Line, then Click NEXT.
3. Click NEXT to insert the titles and axes.

e. Pie Graph:

The pie-Graph is made of a circle divided into wedges, each wedge of some percentage equal to the percentage of the data in the sample. So to graph the pie-chart we need to find the percentage of each wedge and the degree of its angle as follows:

Degrees= (f/n) 360, Percentage % = (f/n) 100.

Here we will apply Pie-graph to Example-1 the non-numerical problem:

Class	Frequency (f)	Degrees=(f/n)360	%= (f/n)100
A	6	72	20%
B	9	108	30%
O	10	120	33%
AB	5	60	17%
	$\sum = n = 30$	$\sum = 360$	$\sum = 100\%$

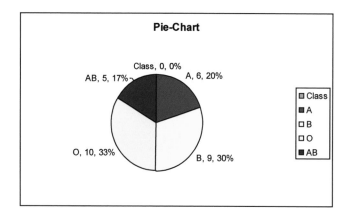

The Pie-Chart was constructed using EXCEL in the following steps:
1. Enter Class in one column, and frequency in the column next to it.

2. Highlight both columns, Click on Insert then Chart, then chose pie, then Click NEXT. Click NEXT to inset the titles and axes, and percentage.

3. If you want to change the title, click on the title.

TI-83+ or TI-84+

Using TI-83+ or TI—84+ graph the Histogram, Frequency polygon, and O-Give for the data from example-6:

a. Histogram:

1. Clear your calculator first as follows: Press 2nd, (+) →7→ 1→ 2, enter.
2. Enter midterm data X_m on L_1, and f on L_2.
3. Set up widows: X-min $\leq$ LV, X-max $\geq$ HV, X-Scale = W.
   Y-min $\leq$ Lf, y-max $\geq$ HV, keep the y-scale 1.
4. 2nd STAT-PLOT enter, Histogram-Enter- L_1 enter, L_2 enter.
5. Graph.

b. Frequency-Polygon:

1. Follow the above steps from 1-4.
2. 2nd STAT-PLOT enter, 2nd graph- L_1 enter, L_2 enter.
3. Graph.

c. O-give

1. Enter Upper-Class Boundaries on L3, frequency on L_4.
2. Change the graph.
3. Change windows.

SPSS

1. Open the Chart Builder from the graphs menu.
2. Click the Gallery tab.
3. Choose Histogram from the list of Chart Types.
4. Drag the Histogram onto the Canvas.
5. Drag a scale Variable to the y-axis drop zone, and click OK.

| Example-9 | In a study of reaction of dogs to a specific stimulus, an animal trainer obtained the following data: |

Class-Limits	Frequency
2.3 – 2.9	1
3.0 – 3.6	3
3.7 – 4.3	4
4.4 – 5.0	16
5.1 – 5.7	14
5.8 – 6.4	4

Use EXCEL to graph a Histogram, Frequency Polygon, and O-give.

Solution:

Graphing a Histogram with EXCEL:

a. Find the class boundaries (± 0.05) then the above table will be:

Class-Boundaries	Freq.
2.25 – 2.95	1
2.95 – 3.65	3
3.65 – 4.35	4
4.35 – 4.95	16
5.05 – 5.75	14
5.75 – 6.45	4

b. Copy the above table on EXCELL Sheet, highlight the 2-columns starting at the top-left-corner and down, then click on Insert → Column of histogram → then on 2D-Column, you will get a Histogram with **gaps-between-the-blocks.**

c. Highlight and copy the graph, then paste on word sheet as shown:

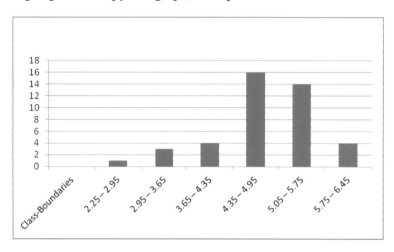

d. Remove the gaps between the blocks as follows: Go back to EXCEL sheet, click on any box then select Format Data Series then move it to zero and the histogram will be like the one shown below, also to add the titles on any side, you click on a title (on the excel sheet) then click on Layout on Chart Tools.

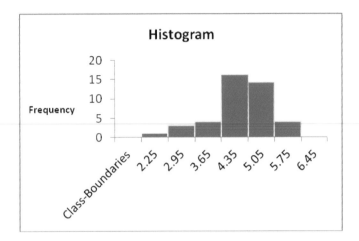

e. To change the size of this graph on this sheet R-click the mouse on any area in this Graph → click on Format Picture then → click on size then change the Height and Width.

Graphing Frequency Polygon with EXCEL:

a. To graph the Frequency Polygon, this requires Class Mid-Points (on x-axis) and Frequency (on y-axis): Mid-Points = On each block (1st value + 2nd value) / 2

Class-Mid Points	Freq.
2.6	1
3.3	3
4	4
4.7	16
5.4	14
6.1	4

Follow the above steps b→e, after choosing Frequency Polygon Graph, as shown

Graphing O-Give with EXCEL:

a. To graph the O-Give we need Class-Boundaries (on x-axis), with Cumulative frequency
    (on y-axis):

Class-Boundaries	C-Freq
2.25	0
2.95	1
3.65	4
4.35	8
5.05	24
5.75	38
6.45	42

Example-10   In a study of 100-women, the numbers shown here indicate the major
reason why each woman surveyed worked outside the house.
Use EXCEL to construct the Pie-Graph

Reason	Number of Women
To support self/family	62
For extra money	18
For something different to do	12
Other	8

Solution:

a. Add to the table above 2-more columns one for the %= (f/n) x100%, and the other
    column for Degrees= (f/n)360, as follows:

$$\% = f/n \times 100\%, \qquad Degrees = f/n \times 360$$

Reason	#of Women	%	Degrees
To Support	62	62%	223.2
Extra Money	18	18%	64.8
Different Things	12	12%	43.2
Other	8	8%	28.8
	$\Sigma f = 100 = n$	$\Sigma = 100\%$	$\Sigma = 360$

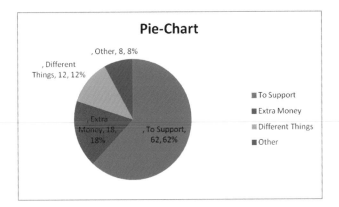

a. To change the size R-click inside the graph, and then click on Format pictures then size.
b. To add the labels on each wedge: R-click inside the graph (on EXCEL sheet) then on Format Data Label and fill up Label Options.

2.3 Practice Problems

1. Construct a Histogram (with 7-classes), frequency Polygon, and O-give for the given Data:

Data: 65 36 49 84 79 56 28 43 67 36 82 22 62 55 72 68
40 37 28 43 88 50 60 56 57 46 39 57 73 65 45 56
75 40 51 70 74 76 48 59 35 62 52 63 32 80 64 53
74 34 51 35 44 45 54 51 55 48 60 76 21 35 61 45
33 61 77 60 85 68 47 52 68 52 69 42 67 34 53 45
62 65 55 61 73 50 53 59 41 54 70 38 50 47 35 26
58 82 74 41.

2. In a general hospital the number of each operation performed in one year is Listed below. Construct the Pie-Chart for the given data:

Type of Operation	Number of cases
Theoretic	20
Bones and Joints	45
Eye, ear, and throat	58
General	95
Abdominal	98
Urologic	74
Proctologic	65
Neurosurgery	23

3. Using 7-classes, construct a Histogram, Frequency Polygon, and O-give for the Given (58) Data:

Data: 66.4 69.2 70.0 71.0 71.9 74.2 74.5 72.1 71.1 70.1 69.3 67.7
    68.0 69.3 70.2 71.2 72.2 75.3 68.0 69.5 70.3 71.3 72.3 68.3
    69.5 70.3 71.3 72.4 68.4 69.6 70.4 71.5 72.6 68.6 69.7 70.5
    71.6 72.7 68.8 69.8 70.6 71.6 72.9 68.9 69.8 70.6 71.7 73.1
    69.0 69.9 70.8 71.8 73.3 69.1 70.0 70.9 71.8 73.5

3 Measures of Data

Measures of Data

In chapte-2 we managed to organize the raw data into a frequency distribution, and then present it by using several graphs. In this chapter we will present the statistical measures that can be used to summarize the data, the most familiar of these measures are:

1. Measure of Average.
2. Measure of Variation.
3. Measure of Position.

3.1 Measure of Average

The measures of Average are:

a. Mean.
b. Median
c. Mode
d. Midrange

a. Mean also known as the arithmetic average, is denoted by $\overline{X}$, read X-bar.

   And is equal to the sum of all data divided by total number of the data
   In the sample, formula given as:

$$\text{Mean for a sample: } \overline{X} = \frac{X_1 + X_2 + X_3 + \ldots}{n} = \frac{\sum X}{n} \quad \ldots\ldots (1)$$

$$\text{Mean for population: } \mu = \frac{X_1 + X_2 + X_3 + \ldots}{N} = \frac{\sum X}{N} \quad \ldots\ldots (2)$$

Where:$X_1, X_2,..$, represents the data , n=sample size, and N=population size

b. Median (MD): Is the midpoint of the data array arranged in order.

c. Mode: Data that occurs most often.

d. Midrange (MR): (HV+LV)/2.

Finding the measures of Average for ungrouped data

Example-1   The point average (GPA) for top-25-ranked medical schools is listed below:
Data: 3.80 3.86 3.83 3.78 3.75 3.64 3.78 3.73 3.64 3.66 3.67
3.57 3.70 3.74 3.70 3.77 3.76 3.68 3.67 3.74 3.65 3.73
3.74 3.70 3.80

Step-1: Arrange the data in ascending order:

Data in ascending order: 3.57 3.64 3.64 3.65 3.66 3.67 3.67 3.68
3.70 3.70 3.70 3.73 3.73 3.74 3.74 3.74
3.75 3.76 3.77 3.78 3.78 3.80 3.80 3.83 3.86

Step-2: Use formula (1) to find the average of the data

Mean for a sample: $X = \dfrac{\sum X}{n} = \dfrac{3.75 + 3.64 + \ldots}{25} = \dfrac{93.09}{25} = 3.7236 \approx 3.724$

Since n=25 which is an odd-number then the median is the data in the middle of the list.

Median (MD) for the odd number of data = 3.73

If we remove one data from the above list, say (3.73) the number of the data n=24 even
Then the median is the average of the two-data in the middle or:
Median of even data = (3.73+3.86)/2 = 3.795.

Mode = 3.70, and 3.74 both were repeated 3-times.
Midrange = (3.57 + 3.86) / 2 = 3.715.

Example-2   Find the mean, median, mode and midrange for the following seven data:
12 9 3 12 14 15 10

Solution: 1. We arrange the data in order: 3 9 10 12 12 14 15

Data in the middle

2. The average of the data $= \dfrac{\sum X}{n} = \dfrac{3+9+10+12+12+14+15}{7} = 10.7$

Note: we round up the average to one digit behind the decimal

3. The median is the data in the middle as shown above = 12

4. Mode of the data is the one that is repeated = 12

    3 9 10 12 12 14 15

5. The midrange is ½ of the sum of the lowest value (LV) and the highest value (HV):

    3 9 10 12 12 14 15

    (15+3)/2 = 9

 Example-3   Find the mean, median, mode, and midrange for the following eight data:

    11 2 4 12 13 15 12 11

Solution: 1. We arrange the data in order: 2 4 11 11 12 12 13 15

    ↑

    Data in the middle

2. The average of the data $= \dfrac{\sum X}{n} = \dfrac{2+4+11+11+12+13+15}{8} = 8.5$

4. The median is the average of the 2-data in the middle $= \dfrac{11+12}{2} = 11.5$

5. Mode of the data is the one that is repeated = 11, and 12

    2 4 11 11 12 12 13 15

5. The midrange is ½ of the sum of the lowest value (LV) and the highest value (HV):

    2 4 11 11 12 12 13 15

    (15+2)/2 = 8.5

Finding the measures of Average for grouped data

Example-4   Here we will use our example-3 for grouped data and try to find the mean. The formula of the mean for a grouped data is:

$$\bar{X} = \frac{\sum f . X_m}{n} \quad ....(3)$$

Class-limits	Class-Boundaries	Frequency (f)	Mid-point X_m	f. X_m
15 – 19	14.5 – 19.5	3	17	51
20 – 24	19.5 – 24.5	18	22	396
25 – 29	24.5 – 29.5	18	27	486
30 – 34	29.5 – 34.5	8	32	256
35 - 39	34.5 – 39.5	3	37	111
		$\sum = n = 50$		$\sum = 1300$

Solution: Using the frequency f and the midpoint X_m from the table, we can find the average (mean) of the grouped data using formula(3)as follows:

$$\overline{X} = \frac{\sum f . X_m}{n} = \frac{1300}{50} = 26$$

The Mode for the grouped data is called Modal class = the class with highest frequency.
For the above grouped data:

Class-limits	Frequency (f)
15 – 19	3
20 – 24	18
25 – 29	18
30 – 34	8
35 - 39	3

Two Modal Classes

Arrows are pointing at **two-Modal classes** {20-24}, and {25-29} with the highest Frequency of 18.

Weighted Mean:

To find the average point of the (GPA) the formula is:

$$\overline{X} = \frac{\sum w.x}{\sum w} \quad \ldots\ldots\ldots(4)$$

Example-5   A college student registered for 4 classes in fall, and received the following grades (on scale 4):

Class	Grade	x-points	w-Credits
Math	A	4	3
Psychology	C	2	3
Biology-I	B	3	4
Chemistry-I	D	1	2

Solution: To find his GPA, student has to use the above formula In the following way:

Class	Grade	x-points	w-Credits	w.x
Math	A	4	3	12
Psychology	C	2	3	6
Biology-I	B	3	4	12
Chemistry-I	D	1	2	2
			$\sum = 12$	$\sum = 32$

Then, GPA $= \overline{X} = \dfrac{32}{12} = 2.7$

The grade point average is 2.7 out of scale of 4.

3.2. Practice Problems (Measure of Average)

1. The ages of 30 women at the birth of their first child was obtained from a general Hospital:

Data: 16 24 16 28 22 17 21 21 22 24 29 20 42 22 22
  22 15   20 21 18   24 23 25   21 23 24 42 22 22 17

Find the following:

a. Mean age of the women.
b. Median age of the women.
c. Mode.
d. Midrange.

2. Find the mean and Modal-class for the given 40- Grouped-Data:

Data: 22 21 37 19 22 22 12 27 16 26 18 25 23 21 15 22 22
   18 23   8 18 17 20   21 16 20 15 17 17 20 19 24 18 25
   30 23 16 26 22 20.

3. Find the GPA for a student who took the following classes with the given Scores:

Class	Grade	x-points	w-Credits
English 101	A	4	3
Psychology	B	3	3
Biology-I	C	2	4
Physiology	C	2	2
Exercise	A	4	3

3.3 Measure of Variation

The measures of variation for samples are:
1. Range. = R
2. Variance = S^2
3. Standard deviation = S

1. Range(R): This was used before in frequency distribution.

$R = HV - LV$

2. Finding the measures of variation for un-grouped data

| Example-6 | Find the variance and standard deviation for the given data.

Data: 21 33 27 35 31 28 26 32 37 30

The formula for the variance for a sample is denoted by S, and is given as:

$$S^2 = \frac{\sum (x - \bar{x})^2}{n - 1} \quad \dots (5) \quad \text{Where, n=sample size, and } \bar{x} \text{ is the sample mean}$$

From the given data we have:

n = number of data = 10.

$\sum X$ = sum of all the data = 21+33+27+35+31+28+26+32+37+30 = 300

Mean for a sample: $\bar{X} = \dfrac{\sum X}{n} = \dfrac{300}{10} = 30$

To find the variance using formula (5) it is easier to construct the following table:

Data (X)	$X - \overline{X}$	$(X - \overline{X})^2$
21	21-30=-9	$(-9)^2 = 81$
33	33-3-=3	$(3)^2 = 9$
27	27-30=-3	$(-3)^2 = 9$
35	35-30=5	$(5)^2 = 25$
31	31-30=1	$(1)^2 = 1$
28	28-30=-2	$(-2)^2 = 4$
26	26-30=-4	$(-4)^2 = 16$
32	32-30=2	$(2)^2 = 4$
37	37-30=7	$(7)^2 = 49$
30	30-30=0	$0^2 = 0$
		$\sum = 198$

Applying Formula (5) gives $S^2 = 22$
And the Standard deviation for the sample S is:

$$S = \sqrt{S^2} = \sqrt{22} = 4.69$$

Note: If the data are taken from a Population instead of sample then:

The variance is $\sigma^2 = \dfrac{\sum (X - \mu)^2}{N}$ (4)

Where, μ is the mean of the population, and N is the sample size of the population.

Then, Variance for the population is $= \sigma^2 = \dfrac{198}{10} = 19.8$

And, the Standard Deviation for the population is $\sigma = \sqrt{\sigma^2} = 4.45$

Finding the measures of Variation for grouped data

For grouped data, we refer back again to the data in our example-4, copy the table from Example-4 and add the last column for f. $(X_m)^2$:

Class-limits	Class-Boundaries	Frequency (f)	Mid-point X_m	f. X_m	f. $(X_m)^2$
15 – 19	14.5 – 19.5	3	17	51	867
20 – 24	19.5 – 24.5	18	22	396	8712
25 – 29	24.5 – 29.5	18	27	486	13122
30 – 34	29.5 – 34.5	8	32	256	8192
35 - 39	34.5 – 39.5	3	37	111	4107
		$\sum = n = 50$		$\sum = 1300$	$\sum = 35000$

Here we will use the short cut formula for deviation of Grouped data given as:

$$\text{Short –Cut- Formula } S^2 = \frac{n[\sum f. X^2_m] - (\sum f. X_m)^2}{n(n-1)} \quad .... (6)$$
For sample variance

Substituting the information from the table into formula (6) we get the variance (S^2) for the Grouped data:

$$S^2 = \frac{50\,(35000) - (1300)^2}{50\,(50 - 1)} \approx 24.48979592$$

Then the standard deviation of the grouped data is $= S \approx 4.95$.

Finding the summary statistics using TI-83+, 84+: Follow these steps:
Refer to Example-1:
1. Clear the calculator by pressing: $2^{nd} \rightarrow + \rightarrow 7 \rightarrow 1 \rightarrow 2$.
2. Press STAT → EDIT → enter
3. Enter the data on L_1, then 2nd Quit.
4. Press STST →CALC →1-Var Stat, enter→ enter.
5. The calculator will display the following information:

1-Var Stats		
$\bar{X} = 3.7236$	$\sigma_x = .0661440851$	Med $= 3.73$
$\sum X = 93.09$	$n = 25$	$Q_3 = 3.775$
$\sum X^2 = 346.7393$	min x $= 3.75$	max x $= 3.86$
$S_x = .0675080242$	$Q_1 = 3.67$	

Using EXCEL to find the above summary statistics: follow these steps:
 Refer to Exapmle-1:

1. Type label on cell A1
2. Enter all 25 data from Example-1 on cells A2-A26
3. Chose any cell on the next columns and type: = AVERAGE(A2:A26) enter this will
    Give the average.
    4. Chose another cell and type : = MEDIAN(A2:A26) enter to get the median
    5. In another cell type = MODE (A2:A26) to get the mode.
    6. And for Standard deviation type =STDEV (A2:A26) enter gives the standard deviation.
    7. For Variance type =VAR (A2:A26) enter.
    8. For range type = MAX (A2:A26) – MIN (A2:A26) enter.

The EXCEL sheet will be as follows (Note data are omitted to fit the page)

	A	B
1	Data	
2	3.8	= AVERAGE(A2:A26)
3	3.86	= MEDIAN(A2:A26)
4	3.83	= MODE(A2:A26)
5	3.78	= VAR(A2:A26)
6	3.75	= STDEV(A2:A26)
7	3.64	= MAX(A2:A26) – MIN(A2:A26)
8	3.78	

The output of EXCEL is as shown:

Data		
3.8		
3.86	3.7236	Average
3.83	3.73	Median
3.78	3.7	Mode
3.75	0.067508	SD
3.64	0.004557	Variance
3.78	0.29	Range
3.73		

These steps can be followed for the rest of the examples on this chapter.

3.4 Practice Problems (Measures of Variation)

1. The time (in seconds) spent by students to take a one-hour test in Math for a class of 26 students are as shown:

   Data: 60 45 34 36 56 54 29 57 60 60 50 40 41
   47 48 47 48 55 55 53 59 54 44 43 41 59

   Find the variance and standard deviation for the above ungrouped data. Use 7 classes.

2. For the grouped-data given in the table below, find the variance and standard Deviation, using the short-cut-formula:

Class limits	Frequency (f)
19 – 25	7
26 – 32	12
33- 39	5
40 – 46	6
47 – 53	2
54 – 60	1
61 - 67	2

3.5 Measures of position

Measures of position are:
1. Z-Score = Z.
2. Quartile = Q.
3. Outlier.

---

1. **z-score** also called standard score is computed using the following formulas:

$$z = \frac{x - \mu}{\sigma} \text{ for population, and } z = \frac{x - \bar{x}}{s} \text{ for sample.}$$

2. Quartile: Quartiles are three $Q_1, Q_2,$ and Q_3, they divide the range of the data into 4 parts each represents 25% of the data.

3. Outlier: is extremely low or extremely high data value.

Example-7 Compute the z-score for the following data for mean $\mu=65$ and standard deviation $\sigma = 15$:

a) x= 75
b) x=45
c) x=95

Solution:

a) $z = \dfrac{75 - 65}{15} = 10/15 = 0.67$

b) $z = \dfrac{75 - 65}{15} = -20/15 = -1.33$

c) $z = \dfrac{95 - 6}{15} = 30/15 = 2$

The quartiles can be represented graphically using the box plot as follows:

Example-8 Construct a box plot for the given 10-data:
45, 83, 11, 278, 34, 210, 33, 47, 77, 140

Solution: **1.** Arrange the data in order:
11, 33, 34, 45, 47, 77, 83, 140, 210, 278

The lowest value of the data is LV = 11, and the highest value HV = 278

2. Find the median: since the data =10 (even) then the median is the average of the 2-data in the middle: ½(47+77) = 62 = Q2

$$1, 33, 34, 45, 47, 77, 83, 140, 210, 278$$

Median= 62 = Q_2

3. Split the data into 2-groups (at the median) and find the median of each group, since each Group has odd number of data then median is a single value in the middle:

Median of the first group is: 34 = Q1, and median of the second group = 140 = Q3:

$$11, 33, 34, 45, 47 \qquad 77, 83, 140, 210, 278$$

Q_1= 34 $\qquad\qquad\qquad$ Q_3 = 140

Now we can draw the box:

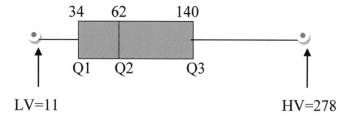

This box was plotted using TI-83 in the following steps:

Graphing Box plot using TI-83 or 84+:
1. Enter the data into L1.
2. Set up the window as follows:
X_{min} = 10 (< LV)
X_{max} = 280 (> HV)
Y_{min} = 0
Y_{max} = 1
3. Press STAT-PLOT →1 → box graph → L1 → frequency =1 enter.
4. Press graph and you'll get the box.
5. To see all the above values press TRACE → at LV, Q1, Q2, Q3, and HV you'll get: 11,34, 62, 140, and 278 respectively.

SPSS for Box Plot:
Graphs menu → open Chart Builder → click Gallery tab → Chart Type → chose Box Plot → Drag the box plot onto the canvas → Drag a scale variable to the y-axis drop zone → Click OK.

Finding the Outlier(s):

| Example-9 | Find the outliers (if any) for the given data: |

Data: 118   123   114   120   119   142   60

| Solution: | 1. Arrange the data in order: n=7 |

60   114   118   119   120   123   142

Median of the data = Q_2

2. Separate the data into 2-groups and find Q1 and Q3:

60   114   118   120   123   142

Q_1 Q_3

3. Find: 1.5(Q3 – Q1) = 1.5(123-114) = 13.5
4. Subtract Q1 – 13.5 = 114–13.5 = 100.5, and
     Q3 + 13.5= 123 + 13.5=136.5

5. Writ the interval with these 2-values, and place the data in the interval, the data that Fall outside this interval will be considered the **outliers**.

These 2- data falls outside the interval

60   [100.5   114   118   119   120   123   **136.5]**   142

All the above data falls inside the interval [100.5 – 136.5] except for 60 and 142 and these are the outliers of the set.

3.6   Practice Problems

For the given data: 105 115 173 208 229 229 230 232 235 238
                      245 245 246 247 249 250 250 254 255 261
       a) Find the quartiles.
       b) Find the outlier(s).

4

Probability & Counting Rules

Probability and Counting Rules

---

*P*robability is the chance of an event to occur. Probability has its own terminology:

- Experiment: is a process that leads to results called outcomes.
- Outcomes: is the result of a single trial.
- Trial: is the number of times the experiment is repeated.
- Sample Space: Is the set of all possible outcomes of experiment.

  Example: Flipping a coin twice.

  Experiment: is the process of flipping the coin.

  Outcomes: are the number of Heads, and Tails that appear.

  Trials: number of times the coin was flipped.

  Sample space: all the Heads and Tails.

There are 3-basic fields of probability:

- Classical Probability.
- Empirical or Relative frequency probability.
- Subjective probability.

4.1   Classical Probability

Classical probability treats certain outcomes as equally likely. In classical probability the sample space is used to determine the numerical value of the probability, and a formula is used as follows:

$$P(E) = \frac{n(E)}{n(s)} = \frac{number\ of\ outcomes\ in\ the\ event\ E}{number\ of\ oucomes\ in\ the\ sample\ s} \qquad (1)$$

Where: E = event, n (E) = number of outcomes in the event E.
S = sample space, n(s) = number of outcomes in the sample space S.

Rounding rules for probability: Probability value should be rounded to 2 or 3 decimal places. Probability can be expressed as fractions, decimals, or percentages. There are 3-typical experiments that are used in probability:

    a. Coins.

    b. Dice

    c. Cards

a. Coins and outcomes:

Example-1: Flip a coin once.

$S = \{H, T\}$, $n(S) = 2^1$, number of trials = 1.

Example-2: Flip a coin twice:

$S = \{HH, HT, TH, TT\}$
$n(S) = 2^2 = 4$. Number of trials = 2

Example-3: Flip coin 3-times:

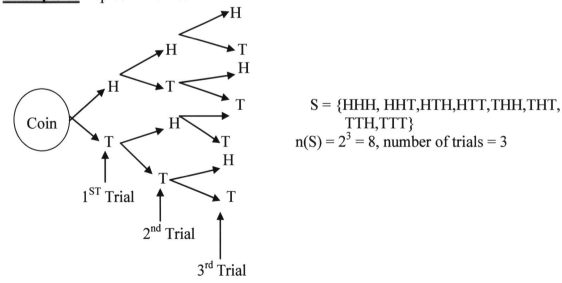

$S = \{HHH, HHT, HTH, HTT, THH, THT,$
$TTH, TTT\}$
$n(S) = 2^3 = 8$, number of trials = 3

b. Dice and outcomes:

Example-4: Rolling a die once.

S = {1, 2, 3, 4, 5, 6}
$n(S) = 6^1$. Number of trials = 1

Example-5: Rolling a die twice. This can be done in table or Tree-Diagram:

	1	2	3	4	5	6
1	11	12	13	14	15	16
2	21	22	23	24	25	26
3	31	32	33	34	35	36
4	41	42	43	44	45	46
5	51	52	53	54	55	56
6	61	62	63	64	65	66

$n(S) = 36$

S= {11, 12, 13, 14, 15, 16; 21, 22, 23, 24, 25, 26; 31, 32, 33, 34, 35, 36;
    41, 42, 43, 44, 45, 46; 51, 52, 53, 54, 55, 56; 61, 62, 63, 64, 65, 66}
$n(S) = 6^2 = 36$, number of trials = 2

Tree –Diagram for rolling a die once:

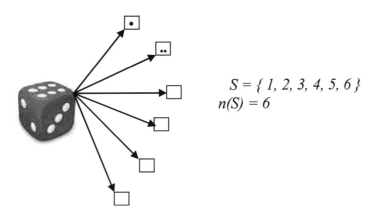

$S = \{ 1, 2, 3, 4, 5, 6 \}$
$n(S) = 6$

Tree-Diagram for rolling a die twice :

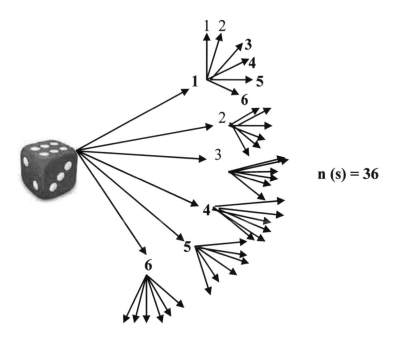

n (s) = 36

c. Cards and outcomes:

n(s) = 52

The sample space for cards is made of 4 suits = {♣ clubs, ♥ hearts, ♠ spades, ♦ diamonds}.
Each suit = 13 cards:
n(clubs) = 13 cards.
n(hearts)= 13 cards.
n(spades) = 13 cards.
n(diamonds) = 13 cards.
The total sample n(S) = 13+13+13+13 = 52 cards.

4.2 Rules of Classical Probability

1. Rules of addition (union):

This is represented by the word "OR" and symbolized by ∪ which means union, or addition of outcomes. The rule of union applies in two cases:

- When two events are mutually exclusive, the probability of event A or event B to occur is equal to the sum of the probability of the 2-events.
- When two events are mutually not exclusive (inclusive) then the probability of event A OR event B is equal to the sum of the 2-events A and B subtracting the probability of the intersected event.

- $P(A \cup B) = P(A) + P(B)$ If A and B are mutually exclusive . . . (2)
- $P(A \cup B) = P(A) + P(B) - P(A \cap B)$ If A and B are inclusive . . . (3)

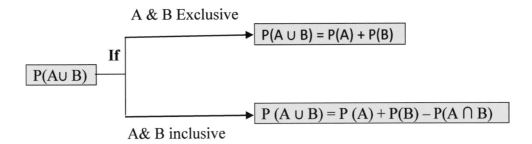

A & B Exclusive

If

$P(A \cup B)$

$P(A \cup B) = P(A) + P(B)$

A & B inclusive

$P(A \cup B) = P(A) + P(B) - P(A \cap B)$

The union can be illustrated by Venn-Diagram as shown:

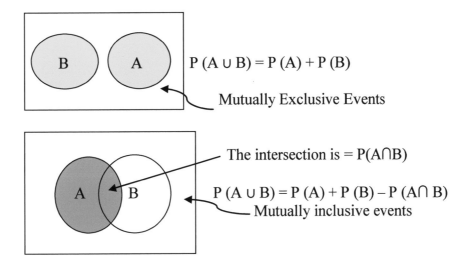

$P(A \cup B) = P(A) + P(B)$

Mutually Exclusive Events

The intersection is $= P(A \cap B)$

$P(A \cup B) = P(A) + P(B) - P(A \cap B)$
Mutually inclusive events

Solving Classical Probability Problems:

The easiest way to solve the classical probability is to prepare the sample space first then use the right formula. Here we will give one example of each of the three classical experiments: Coins, Dice, and Cards.

Example-6

A coin is flipped 3-times. Find the probability of getting 2-heads.

Solution

Use Tree-Diagram to write the sample space first:

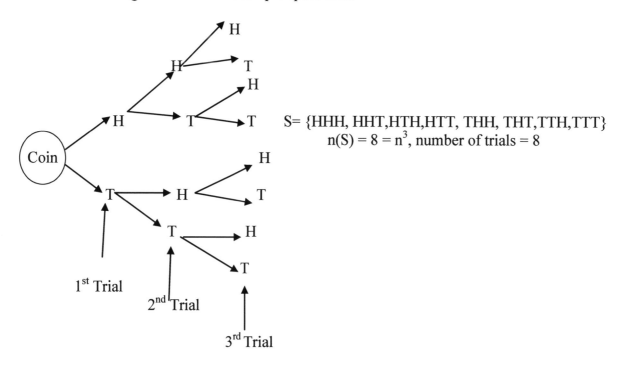

S= {HHH, HHT,HTH,HTT, THH, THT,TTH,TTT}
n(S) = 8 = n^3, number of trials = 8

The event in question is E=2H, and n(E) = {HHT,HTH,THH} = 3 .

To find probability of the event use the first rule: $P(E) = p(2H) = \dfrac{n(2H)}{n(S)} = \dfrac{3}{8}$

Example-7

A die is rolled twice. Find the following probabilities:

a. A sum of 7.
b. A sum of 6 or 9.
c. A sum less than 8.
d. Doubles.
e. Doubles or sum of 4.

Solution:

First we will show the sample space in table form:

	1	2	3	4	5	6
1	11	12	13	14	15	16
2	21	22	23	24	25	26
3	31	32	33	34	35	36
4	41	42	43	44	45	46
5	51	52	53	54	55	56
6	61	62	63	64	65	66

$n(S) = 36$

a. Probability of sum of 7 = p(Σ=7) = $\dfrac{n(\Sigma=7)}{n(S)} = \dfrac{n(61,52,43,34,25,16)}{n(s)} = \dfrac{6}{36} = \dfrac{1}{6}$

b. Probability of sum of 6 or 9 = p{Σ (6 U 9)} since the two events are mutually inclusive we will use rule – 2.

	1	2	3	4	5	6
1	11	12	13	14	15	16
2	21	22	23	24	25	26
3	31	32	33	34	35	36
4	41	42	43	44	45	46
5	51	52	53	54	55	56
6	61	62	63	64	65	66

$n(S) = 36$

P{Σ (6 U 9)} = P(Σ = 6) + P(Σ = 9)

$$= \frac{n(51,42,33,24,15)}{n(S)} + \frac{n(63,54,45,36)}{n(S)} = \frac{5}{36} + \frac{4}{36} = \frac{9}{36} = 25$$

c. A sum less than 8, this is a single event, we use formula-1. E= (Σ <8)

	1	2	3	4	5	6
1	11	12	13	14	15	16
2	21	22	23	24	25	26
3	31	32	33	34	35	36
4	41	42	43	44	45	46
5	51	52	53	54	55	56
6	61	62	63	64	65	66

n(S) = 36

Then P(Σ < 8) = $\dfrac{n(61,51.52,41,42,43,31,32,33,34,21,22,23,24,25,11,12,13,14,15,16)}{n(S)}$

$$=\frac{21}{36}=\frac{7}{12}$$

d. Probability of (doubles)

	1	2	3	4	5	6
1	11	12	13	14	15	16
2	21	22	23	24	25	26
3	31	32	33	34	35	36
4	41	42	43	44	45	46
5	51	52	53	54	55	56
6	61	62	63	64	65	66

n(S) = 36

Then P (doubles) = $\dfrac{n(doubles)}{n(S)}$ = $\dfrac{n(11.22,33,44,55,66)}{n(S)}$ = $\dfrac{6}{36}=\dfrac{1}{6}$

e. Probability of doubles or sum of 4.

	1	2	3	4	5	6
1	11	12	13	14	15	16
2	21	22	23	24	25	26
3	31	32	33	34	35	36
4	41	42	43	44	45	46
5	51	52	53	54	55	56
6	61	62	63	64	65	66

Point of intersection
p(doubles ∩ Σ=4)

n(S) = 36

Since these events are mutually inclusive then we have to use formula-3.

$P(\text{doubles} \cup \Sigma = 4) = p(\text{doubles}) + p(\Sigma = 4) - p(\text{doubles} \cap \Sigma = 4)$

$$= \frac{6}{36} + \frac{3}{36} - \frac{1}{36} = \frac{8}{36} = \frac{2}{9}$$

Example-8

A card is drawn from a deck of 52 cards. Find the probabilities of getting the Following events:

a. Getting a 10
b. Getting an Ace.
c. Getting a face.
d. Getting a king.
e. Getting a queen or a face.
f. Getting a face or a red card.

The sample space for cards is shown below: Cards are made of 4-suits:

Hearts (♥), Diamonds (♦) Clubs (♣), and Spades (♠). Each suit = 13 cards, two suits Are red in color, and two black in color. The sample space n(s) = 52

A	2	3	4	5	6	7	8	8	9	10	J	Q	K	= 13	26 Red
A	2	3	4	5	6	7	8	8	9	10	J	Q	K	= 13	
A	2	3	4	5	6	7	8	8	9	10	J	Q	K	= 13	26 Black
A	2	3	4	5	6	7	8	8	9	10	J	Q	K	= 13	n (S) =52

A's means: 1. Faces are: Queen (Q), Jack (J), and King (k), n(faces) = 12, n(red) = 26, n (black) = 26, n(each suit) = 13.

 ⟸ n (A's) = 4

 n(Faces) = 12

Example-8

a. $P(10) = \dfrac{n(10)}{n(s)} = \dfrac{4}{52} = \dfrac{1}{13}$

b. $P(A's) = \dfrac{n(A's)}{n(S)} = \dfrac{4}{52} = \dfrac{1}{13}$

c. $P(Face) = \dfrac{n(Face)}{n(S)} = \dfrac{12}{52} = \dfrac{3}{13}$

d. $P(K) = \dfrac{n(K)}{n(S)} = \dfrac{4}{52} \dfrac{1}{13}$

e. P(Q U F) the events here are mutually inclusive then:

union

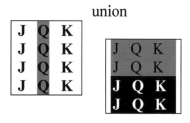

P (Q U F) = P (Q) + P (F) − P (Q ∩ F)

$= \dfrac{4}{52} + \dfrac{12}{52} - \dfrac{4}{52} = \dfrac{12}{52} = \dfrac{3}{13}$

f. P (F U R) this is inclusive events = P(F) + P(R) − P(F∩ R) = $\dfrac{12}{52} + \dfrac{26}{52} - \dfrac{6}{52}$

$= \dfrac{32}{52} = \dfrac{8}{13}$

A	2	3	4	5	6	7	8	8	9	10	J	Q	K
A	2	3	4	5	6	7	8	8	9	10	J	Q	K
A	2	3	4	5	6	7	8	8	9	10	J	Q	K
A	2	3	4	5	6	7	8	8	9	10	J	Q	K

n(R)= 26
n(F) = 12

2. Rules of Multiplication :

This uses the word AND symbolized by ∩ which means intersection, or multiplication of outcomes. The rule of multiplication applies in two cases:

- When two events are independent on each other, the probability of event A and event B is equal to the product of the probability of the 2-events.
- When two events dependent on each other, then the probability of event A and event B is equal to the product of the 2-events A and B assuming that the first event has already occurred.

- $P(A \cap B) = P(A).P(B)$ If A, and B are independent .... (4)
- $P(A \cap B) = P(A).P(B/A)$ if A, and B are dependent. ... (5)

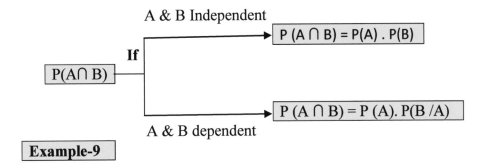

Example-9

Two cards are drawn from a deck of 52-cards **without replacement.**
Find the following probabilities:

a. Getting 2-queens.
b. Getting a diamond, and a spade.
c. Getting black card, and a red card.

Solution:

a. $P \text{ (getting 2-queens)} = P(Q_1 \cap Q_2) = P(Q_1) . P(Q_2/Q_1)$
$$= 4/52. \ 3/51 = 1/221$$

b. $P \text{ (diamond and spade)} = P(\blacklozenge \cap \spadesuit) = P(\blacklozenge) . (\spadesuit / \blacklozenge) =$
$$= (13/52)(13/51) = 13/204$$

c. $P \text{ (black and red)} = P(\text{black}) . P(\text{red/black}) = (26/52)(26/51) = 13/51$

Example-10

Redo Example (9)but **with replacement:**

   a. P (getting 2-queens) = P ($Q_1 \cap Q_2$) = P (Q_1). P(Q_2)
                         = (4/52) (4/52) = 1/169

   b. P (diamond **and** spade) = P (♦ $\cap$♠) = P (♦). (♠)=
                         = (13/52) (13/52) = 1/16

   c. P (black **and** red) = P (black).P (red) = (26/52) (26/52) = ¼

Probability with Tree Diagram

Example-11

A box contains 7-balls, 4 red and 3 blue. Two balls are drawn from the box without
Replacement:
a. Find the probability of selecting a blue ball in the second trial.
b. Find the probabilities of all the outcomes.

Solution:

Drawing a tree diagram for the problem makes it easier to solve.

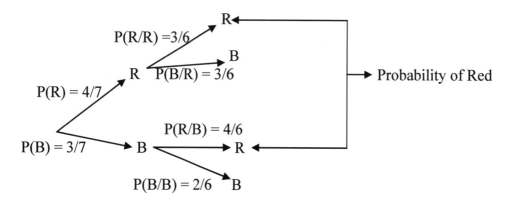

   a. P (B) = (4/7)(3/6) + (3/7)(2/6) = 18/42 = 3/7.

   b. P (all the outcomes) = P (RR or RB or BR or BB)

$$= P \text{ (RR)} \cup P \text{ (RB)} \cup P \text{ (BR)} \cup P \text{ (BB)}$$

$$= P(R) .P(R) + P(R).P (B) + P (B) .P(R) + P (B) .P (B)$$

Then, P (all the outcomes) = (4/7) (3/6) + (4/7) (3/6) + (3/7) (4/6) + (3/7) (2/6) =1
This is the total Probabilities of the outcomes in the sample space.

Other Rules of Multiplication:

Other rules are used in multiplication such as:

- Factorial
- Permutation
- Combination

Factorial: Is symbolized as ! for example n! = n (n-1) (n-2) . . is the product of integers
From 1 to n

Example-12

$$7! = 7.\ 6.\ 5.\ \ 4.\ 3.\ \ 2.\ 1$$

Permutation: Is symbolized as nPr or P(n,r) means taking n-distinct elements r at a time.
Permutation is used **when order is important.** The formula used for permutation is:

$$\text{Permutation} = nPr = p(n,r) = \frac{n!}{(n-r)!} = \frac{n(n-1)(n-2) \ldots 1}{(n-r)!}$$

Example-13

In how many ways can the letters {a, b, c, d} be lined up taking 2 at a time, without-
Replacement?
Answer: n = 4, r = 2 then,

$$\text{Number of ways} = P\ (4,2) = \frac{4!}{(4-2)!} = \frac{4.3.2!}{2!} = 12$$

Note: 0! = 1, and 1! = 1

$$nPr = n(n-1)\ (n-2)\ldots\ (n-r+1) = \frac{n!}{(n-r)!}$$

$$nPn = n(n-1)\ldots 3\ .\ 2.1$$

Example-14

In how many ways can a department select a Director, a chair from 8 full time faculties,
If sharing is not allowed?
Answer: Here we see the order is important, then the number of ways = 8 P 2 = 56

Example-15

If 7-cars are in a car race, in how many ways can 4 of them finish 1st, 2nd, third, and 4th?
Answer: since order is important in this question then:

$$\text{The number of ways} = 7P4 = P(7,4) = \frac{7!}{(7-4)!} = 840$$

Using TI: 1. Enter 7
2. MATH → PRB → 2 →4→ enter

Combination: Is symbolized by n C r or C (n,r), and consider the set of n-distinct elements, a combination of them is a subset, or unordered sub collection. Combination is used when **order is not important.** The formula used for combination is:

$$\text{Combination} = nCr = C(n,r) = \frac{n!}{(n-r)!r!} = \frac{n(n-1)(n-2)\ldots 1}{(n-r)!\ r!}$$

Example-16

In how many ways can a faculty chose 3 students out of 10 for math club help?
Answer: Here we see the order is not important, then
The number of ways = 10 C 3 = C(8, 3) = 120

Using TI: 1. Enter 10
2. MATH → PRB → 3 →3→ enter

Example-17

To see the difference between nPr, and nCr find both taking n=9, and r=5.
Answer: nPr = 9P5 = 15120. And nCr = 9C5 = 126.

Combination, Binomial, and Pascal:

Combination C(n,r) is very important, because it is used in Binomial Probability as the coefficient of the formula:

$$P(r) = C(n,r)\, p^r\, q^{n-r} = \frac{n!}{(n-r)!\ r!} p^r\, q^{n-r}$$

Also, this formula is used in the Binomial Theorem of Algebra:

$$(x + y)^n = \sum_{r=0}^{n} C(n, r) \, x^r \, y^{n-r}$$

Example-18

Compute: $(y + x)^3 = C(3,0) \, y^0 \, x^3 + C(3,1) \, y^1 \, x^2 + C(3,2) \, y^2 \, x^1 + C(3.3) \, y^3 \, x^0$

$$= \quad 1 \qquad\qquad 3 \qquad\qquad 3 \qquad\qquad 1$$

Then, $\quad y + x)^3 = x^3 + 3y \, x^2 + 3 \, y^2 \, x + y^3$

So student can find a binomial of any power using this simple combination.
Pascal also has used Combination in his Triangle, as shown:

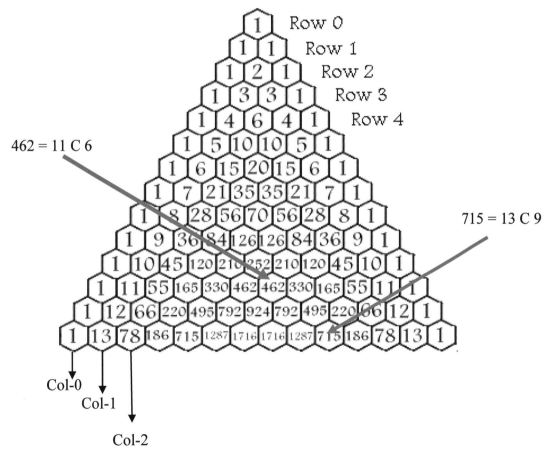

The cell with number (462) is located at Row-11(starting with 0 from the top), and
Column-6 (starting with 0 from left).
Similarly, the cell with number (715) is located at Row-13(starting with 0 from the top), and
Column-6 (starting with 0 from left).

4.3 Conditional Probability

In solving conditional probability we have to consider the other event that has occurred before. Tree-diagrams, tables, and Venn-Diagrams, illustrates conditional probability and make it easier to solve.

Conditional Probability: Is the chance of an event to occur being affected
By another event that has already happened before.

Example- 12

A box contains 6-balls: 4-white, and 2-green. Two balls are drawn one at a time **Without replacement.** Find the probability that white and green balls are drawn.

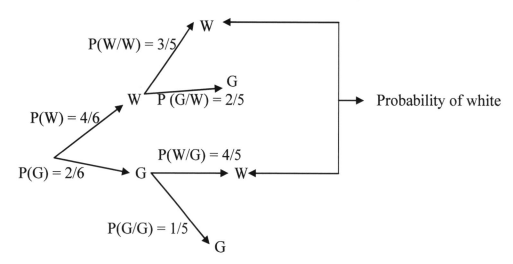

Find P (white and green) = P (W ∩ G) ∪ P (G ∩ W) which is represented by two paths as shown above

$$= P (W) P (W/W) + P (G) P (G/W)$$

$$= (4/6) (3/5) + (2/6) (4/5)$$

$$= 2/5 + 4/25 = 14/25 = .56$$

Example- 13

In stock market, stocks go up and down, and stockbrokers use research or their
Feelings to find out: when the stocks are going up. A brokerage of a firm conducted
A study on 100- brokers to find out which one of the two methods is the best, using
The results shown on the table:

Method used	Stocks up = SU	Stocks down = SD	Total
Research = R	28	17	45
Feelings = F	28	27	55
Total	56	44	n=100

Solution:

1. Find the following probabilities:

a) P (SU) = ΣSU / n(S) = 56/100 = 0.56

Stocks up = SU
28
28
56

b) P (SD) = 1- P(SU) OR ΣSD / n(S) =44/100 = 0.44

Stocks down = SD
17
27
44

c) P(R) = ΣR / n(S) = 45/100 = 0.45

Research = R	28	17	45

d) P(F) = ΣF / n(S) = 55/100 = 0.55

Feelings = F	28	27	55

e) P(R ∩ SU) = 28/ 100

Method used	Stocks up = SU	Stocks down = SD	Total
Research = R	28	17	45
Feelings = F	28	27	55
Total	56	44	n=100

f) P (F ∩ SD) = 27/100

Method used	Stocks up = SU	Stocks down = SD	Total
Research = R	28	17	45
Feelings = F	28	27	55
Total	56	44	n=100

2. What is the probability that a person who uses research picks stocks that go up?
This can be stated as: Find the probability that a broker picked the stocks that go up, given that he used research. This is a conditional probability represented symbolically as:

$$P(SU/R) = \frac{P(SU \cap R)}{P(R)} = \frac{28/100}{45/100} = 28/45 = 0.62$$

Venn-Diagram:

The above problem can be represented in Venn-Diagram as follows:

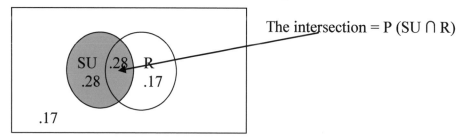

The intersection = $P(SU \cap R)$

Bayes' Theorem:

Bayes 'Theorem was formulated by reverend Thomas Bayes (1761). The theorem deals with conditional probability, and is used to find $P(A/B)$ when the available information is not immediately compatible with that required to apply the definition of conditional probability directly.

Example- 14

In a research smoke versus cancer, suppose a research was conducted to study the affect of cigarettes on cancer. If the probability that a person gets lunch cancer (C), given that he smokes one or more pack of cigarettes(S) a day is (0.1) or in symbols → $P(C/S) = 0.1$, $P(S) = 0.2$, $P(S') = 0.8$, and $P(C/S') = 0.3$, where S' means does not smoke. Find $P(S/C)$.

Solution:

Using the formula we compute $P(S/C)$

$$P(S/C) = \frac{P(S)\, P(C/S)}{P(S)\, P(C/S) + P(S')\, P(C/S')} = \frac{(0.2)(0.1)}{(0.2)(0.1)+(0.8)(0.3)} = \frac{0.02}{0.26} = 0.08$$

Using tree-diagram illustrates this solution easier:

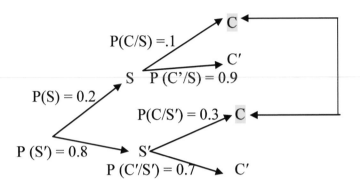

The question is conditional probability:

$$P(S/C) = \frac{P(S)\, P(C/S)}{P(S)\, P(C/S) + P(S')\, P(C/S')} = \frac{(0.2)\,(0.1)}{(0.2)\,(0.1) + (0.8)\,(0.3)} = \frac{0.02}{0.26} = 0.08$$

To find the probability of the subset $P(C) = (0.2)\,(0.1) + (0.8)\,(0.3) = 0.26$ this can be illustrated With Venn-diagram as shown:

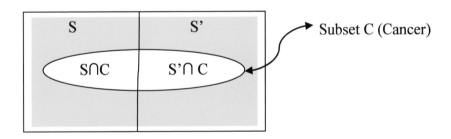

$$P(C) = P(S \cap C) + P(S' \cap C)$$
$$= P(S)\, P(C/S) + P(S')\, P(C/S')$$
$$= (0.2)\,(0.1) + (0.8)\,(0.3)$$
$$= 0.02 + 0.24 = 0.26.$$

4.4 Empirical Probability

Empirical probability depends on actual experience to determine the likelihood of outcomes. The outcomes are determined by the frequency.
Formula used in Empirical probability:

$$P(E) = \frac{Frequency\ of\ the\ class}{Total\ frequancy\ of\ the\ distribution} = \frac{f}{n} \quad \ldots (6)$$

Example- 15

From example-1 on page-7, the blood type problem the following data were given;

Blood Type	Frequency (F)
A	6
B	9
AB	5
O	10
	$\Sigma F = n = 30$

A person was selected at random. Find the probability that the person has:
   a. Type O blood.
   b. Type A or Type AB.
   c. Does not have type B blood.

Solution:

Using rule-1 $P(O) = \dfrac{f}{n} = \dfrac{10}{30} = \dfrac{1}{3}$

   a. The events here are mutually exclusive then : P(A U AB) = P(A) + P(AB)

$$= \frac{6}{30} + \frac{5}{30} = \frac{11}{30}$$

   b. Does not have type B means it is type A or AB or O = P(A) + p(AB) + P(O)

$$= \frac{6}{30} + \frac{5}{30} + \frac{10}{30} = \frac{21}{30}$$

Probability is also expressed as percentage → $\boxed{P(E) = \%}$

Example- 16

If the data in example 15 are given in percentage as follows:

Blood Type	%
A	20%
B	30%
AB	17%
O	33%
	Σ% = 100%

Then the answer to the questions will be:

a. P(O) = 33%

b. P(A U AB) = P(A) + P(AB) = 20% + 17% = 37%

c. P(not B) = P(A) + P(AB) + P(0) = 20% + 17% + 33% = 70%

Probability of the complement:

Not A means complement of A = A¨ = A bar = 1 – A

Using Venn-Diagram to show the complement:

A¨ is the complement of A

$P (A^-) = 1 - P(A)$

$P (A) + P (A^-) = 1$

4.5 Practice Problems

1. A card is drawn from a deck of 52-cards. Find the following probabilities:
    a) A face.
    b) An ace or a spade.
    c) A 9 and a club.
    d) A black card.

2. In a survey: 18 people preferred to take bus to work, 29 people prefer to drive, 13 People prefer to take a train. If a person is selected at random, find the probability That the person preferred to drive.

3. In a women shoe store, during a sale, 17 high heel pair of shoes were sold, 4-pair of
   Flat shoes and 15 pair of sandals. If a customer is selected at random: Find the
   Probability that she bought:
   a) A pair of high heels.
   b) A pair of sandals or, a pair of flat shoes.
   c) A pair that is not flat.

4. A dice is rolled twice; find the probability of
   a) Sum of 6.
   b) Sum less than 4.

5. A survey was conducted on "How many women like to wear makeup?"
   The following results were made:

	Age 18-25	Age 26-45	Age 46-70
Women who likes to wear makeup	18	30	12
Women who does not like to wear makeup	24	12	4

   If a woman is selected at random, find the following probabilities:
   a) The woman likes to wear makeup / given that she belongs to age group of 18-25
   b) The woman does not like to wear makeup, given that she belongs to the group of
      Age 46-70.

6. How many different license plates can be made of 4-digits followed by 3-letters, if:
   a) Repetition is allowed.
   b) Repetition is not allowed.

7. In a small department 15% of people were of type-blood AB. If 5-people are selected
   At random. Find the probability that at least one of them is of type AB-Blood.

8. In a STAT-Class the caps is 25, there is still 5-seats left. How many different ways can
   5-students be selected to fill the cap?

9. Student of major of Mathematics wants to register in classes in fall: He can select 3 courses
   From Mathematics, 2 of Bossiness major, 1 from English. How many different schedules
   Can he make?

11. In how many different ways can 5-players be seated in a raw?

12. In a college with 7 faculty members, and 20 students, there is to be formed a committee of
    2 faculty, and 3 students. How many committees are possible?

5 Probability Distributions

Gorguis/ Statistics Tools / 64

5

Probability Distributions

To Study the nature of Population it would be ideal if we examine its entire elements First, but this is not always possible, because some of the elements are physically inaccessible, or uncertain. For more practical purpose, sufficiently accurate results may be more quickly and inexpressibly obtained by examining only a small part of the parameter called sample. Samples may be collected or selected in different ways. A systematic sample is one selected according to some fixed system. Most statistical techniques use elements of randomness in the sampling. In this chapter we are dealing with random variables. Although the concept of random sample is easy, but sometimes it's not clear how to obtain one.

> **Random Variable:** is a variable whose value is obtained by chance.
> **Discrete Random Variable**: Is the value of the random variable that is
> Determined by observation or in theory.

In chapter-1 we have divided the variables (data) into two groups: Discrete variables, and Continuous variables, in this chapter we will concentrate on the discrete variables. That is variables that have a finite number of possible values or infinite values that can be counted.

5.1 Discrete Probability Distribution:

This section will be concentrated on a variable (X) that will represent the outcomes of the experiment used in the question, and will take discrete values such as: 0, 1, 2, .etc. The experiment used will be the same one presented in probability in chapter-4.

Constructing a Probability Distribution for a Discrete Random Variable:

Example-1

Construct a table and graph for the number of Tails (T) that will occur from flipping coin Four-times. This is an experiment with 4-trials, to write the sample space accurately we will Show the tree-diagram:

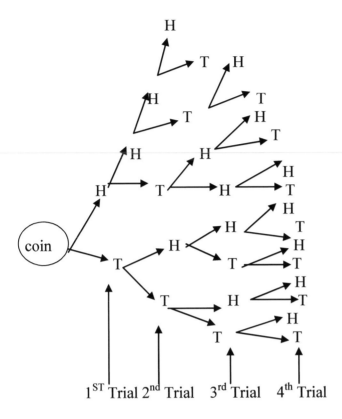

1ST Trial 2nd Trial 3rd Trial 4th Trial

The sample space of the experiment is:
S = {HHHH, HHHT, HHTH, HHTT, HTHH, HTHT, HTTH, HTTT,
THHH, THHT, THTH, THTT, TTHH, TTHT, TTTH, TTTT}

And the sample size $n(S) = 2^4 = 16$ (where 4 is the number of trials).

Solution:
Let the number of Tails (T) that appears = X = 0, 1, 2, 3, 4
The probability of the number of tails that appear = P(X)

The probability of each event is: (with highlighted elements for n(s))
X= T = 0 means zero number of tails →P(X) = n(X) / n(S) = 1/16.

S = {HHHH, HHHT, HHTH, HHTT, HTHH, HTHT, HTTH, HTTT,
THHH, THHT, THTH, THTT, TTHH, TTHT, TTTH, TTTT}

X= T = 1 (with one tail only) →P(X) = n(X) / n(S) = 4/16:

S = {HHHH, HHHT, HHTH, HHTT, HTHH, HTHT, HTTH, HTTT,
THHH, THHT, THTH, THTT, TTHH, TTHT, TTTH, TTTT}

X= T = 2 (with two tails) →P(X) = n(X) / n(S) = 6/16:

S = {HHHH, HHHT, HHTH, HHTT, HTHH, HTHT, HTTH, HTTT,
   THHH, THHT, THTH, THTT, TTHH, TTHT, TTTH, TTTT}

X= T = 3 (with three tails) →P(X) = n(X) / n(S) = 4/16:

S = {HHHH, HHHT, HHTH, HHTT, HTHH, HTHT, HTTH, HTTT,
   THHH, THHT, THTH, THTT, TTHH, TTHT, TTTH, TTTT}

X= T = 4 (with four tails) →P(X) = n(X) / n(S) = 1/16:

S = {HHHH, HHHT, HHTH, HHTT, HTHH, HTHT, HTTH, HTTT,
   THHH, THHT, THTH, THTT, TTHH, TTHT, TTTH, TTTT}

Table and graph of the experiment are:

X	P(X)
0	1/16
1	4/16
2	6/16
3	4/16
4	1/16
	Sum=1

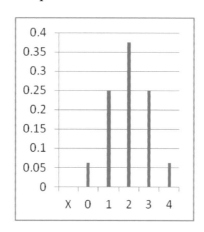

Example-2

In one semester a student registered for 4-classes. The chances that the student will Pass 1, 2, 3, or 4 classes are: 30%, 25%, 20% and 25% respectively. Construct the Probability distribution table and graph for the problem.

Solution:

Let the % of passing = X, with probability = P(X), Then the graph are as shown below:

X	1	2	3	4	
P(X)	30%	25%	20%	25%	ΣP(X) = 1

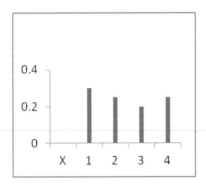

From the above examples we gather the following rules for the discrete probability:

Rules of Discrete Probability:
1. Sum of all the outcomes should add up to 1 → Σ P(X) = 1.
2. Probability of each outcome must be positive → $0 \leq P(x) \leq 1$.

Descriptive Statistics for the Discrete Probability

In this section we will compute the mean of the discrete probability for the population (μ), the variance (σ^2), and the standard deviation (σ) using the following formulas:

Mean $= \mu = \Sigma\, X\, P(X) = E(X)$ $\qquad\qquad$ .... (1)

Variance $= \sigma^2 = [\Sigma X^2\, p(x)] - \mu^2$ $\qquad$ .... (2)

Standard Deviation $= \sigma = \sqrt{[\Sigma X^2\, p(x)] - \mu^2}$ $\quad$ .... (3)

Example-3

Back to our example of flipping coin 4-times. Find the mean (μ), variance (σ^2), and standard deviation (σ).

Solution:

From the given table: To find the mean, variance and standard deviation, we have to add to the table columns for (X^2), (X.P(X)), and (X^2P(X)) as shown on the next table:

X	0	1	2	3	4
P(X)	1/16	4/16	6/16	4/16	1/16

X	P(X)	X.P(X)	X²	X².P(X)
0	1/16	0	0	0
1	4/16	4/16	1	4/16
2	6/16	12/16	4	24/16
3	4/16	12/16	9	36/16
4	1/16	4/16	16	16/16
		Σ X.P(X) = 2		Σ X². P(X) = 5

From the third column: the mean $\mu = \Sigma$ X. P(X) = 2.
From the 3rd and the 4th column we get the variance (σ^2):
$\sigma^2 = \Sigma [X^2 P(X)] - \mu^2 = 5 - (2)^2 = 5 - 4 = 1$, and $\sigma = 1$.

Expectation E(X) is the same concept as the mean. The expectation value of the discrete random variable of probability distribution is the theoretical average of the variables and is measured in the same way as the regular mean using the same formula:

$$E(X) = \mu = \Sigma \text{ X. P(X)} \quad \dots (4)$$

Solving Example-2 using TI-83+ or 84+

1. Enter x-values on L1, P(X) on L2.
2. In L3 type the product of the two values from L1, and L2.
3. In L4 type the product of the square L1 times L2 as shown:

L1	L2	L3	L4
0	1/16	L1* L2	L1² * L2
1	4/16	"	"
2	6/16	"	"
3	4/16	"	"
4	1/16	"	"

4. To find the mean: press 2nd STAT → MATH -→ 5 enter (L1) enter = 2.
5. To find the variance and standard deviation you can follow the same steps as the formula or use the following steps:

   STAT → CALC → 1 enter L1, L2 ENTER, the calculator will display
   The following information: $\bar{X}$ =2, and $\sigma = 1$.

$$
\begin{array}{l}
\overline{X} = 2 \\
\Sigma\, X = 2 \\
\Sigma\, X^2 = 5 \\
\sigma_X = 1 \\
n = 1
\end{array}
$$

5.2 Binomial Probability Distribution

Binomial distribution is used for the experiments with two outcomes such as the experiment of flipping a coin, the outcome is head (H) or tail (T), and the family of children the outcome is Boy (B) or girl (G), and this can be extended to experiment with more than two outcomes, but can be reduced into two such as a test of multiple choice the two outcomes will be using yes or no...etc. The probability of Binomial Distribution must satisfy the following rules:

Rules of Binomial Probability Distribution:
- Each experiment has a fixed number of trials that are independent on each other.
- Each trial has 2-outcomes:
  Success= S with probability = p
  Failure = F with probability = q = 1-p

Formula used for Probability of Binomial Distribution is:

Binomial Probability Distribution Formula:

$$
P(X) = \frac{n!}{(n-X)!\, X!}\; P^X\, q^{\,n-x} \qquad \ldots\ (4)
$$

Where, X= number of success in each trial, $0 \le X \le n$.

Example-4

For a family with four children, find the probability of having 3-girls.

Solution:

The sample for this problem is exactly the same as our previous example of flipping the coin 4-times, that is: Replacing H with B (for the boys) and T with G (for the girls):
S = {BBBB, BBBG, BBGB, BBGG, BGBB, BGBG, BGGB, BGGG,
    GBBB, GBBG, GBGB, GBGG, GGBB, GGBG, GGGB, GGGG}
With n(S) = 16 = 2^4. Then the set of 3G = {BGGG, GBGG, GGBG, GGB}➔ n(3G) = 4

To answer the above question directly from using the sample and classical probability the answer is:

$$P(3G) = \frac{n(G)}{n(S)} = \frac{4}{16} = 1/4 = 25\%$$

Or we can answer the question using formula (4): With p=q=1/2

$$P(X) = \frac{n!}{(n-X)!\, X!}\, P^x\, q^{n-x} = \frac{4!}{(4-3)!\, 3!}\,(1/2)^3\,(1/2)^1 = 4(1/2)^4 = \frac{1}{4} = 25\%$$

Or we can use Binomial probability table as follows:
- Look for n=4 on the first column, and look for X= 3 on the second column next to n.
- Look for probability of success p = ½ on the first raw.
- The intersection between them gives the probability as shown:

n	X	P 0.5
2		
3		
4	0	
	1	
	2	
	3	0.250

Using TI to find the Binomial Probability:

Using the same example-3 with n= 4, p=.5, and X=3, use TI as follows:
2nd DIST → 0 for binompdf (type in here 4, .5, 3) ENTER
Calculator will display 0.25.

Using EXCEL to find the Binomial Probability follow these steps:
- Type X on A1, and P(X) on B1.
- Enter X-data on A2-A6, and enter P(X) on B2-B6 as shown on the table:

X	P(X)
0	=BINOMDIST(0,4,.5,FALSE)
1	=BINOMDIST(1,4,.5,FALSE)
2	=BINOMDIST(2,4,.5,FALSE)
3	=BINOMDIST(3, 4,.5,FALSE)
4	=BINOMDIST(4,4,.5,FALSE)

The result on EXCEL and graph is as shown:

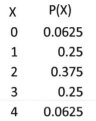

X	P(X)
0	0.0625
1	0.25
2	0.375
3	0.25
4	0.0625

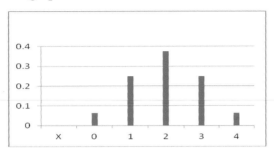

Descriptive Statistics for the Binomial Probability

Formulas :
Mean = μ = n P
Variance = σ^2 = n p q

Standard Deviation = $\sigma = \sqrt{npq}$

Example-5

In example-2 find the mean, variance and standard deviation.
From the experiment we have the following information:
n = number of trials = 4.
X = number of success =3, and P = ½ = q.

Solution:

Mean = μ = nP = 4(1/2) = 2
Variance = $\sqrt{npq}$ = 4(1/2)(1/2) = 1, and Standard Deviation = 1

The Normal Approximation to the Binomial Distribution:

Making direct calculation to compute the probability with binomial distribution could be tedious if the number of trials is large. This problem can be solved using the normal
Model where the distribution can be approximated using the same mean and standard deviation which turns to be a very good approximation. We could find the problem as:

$$P(X < a) = P\left(Z < \frac{X - \mu}{\sigma}\right) = P(Z < b) \qquad \ldots (5)$$

Example-6

Suppose a=1180, n=25000, and P= .05
Then the mean $\mu = np = (25000)(.05) = 1250$, and

Standard deviation $\sigma = \sqrt{npq} = 34.46$.
Then to calculate the probability for X= 1650, using Binomial formula (4) gives:
$P(X = 1180) = {}_{25000}C_{1180} (.05)^{1180} (.95)^{25000-1180}$
(If you try to use TI-83, you will not get any result)
But this can be solved using the normal model (5) as follows:

$$P(X < 1180) = p\left(Z < \frac{1180 - 1250}{34.46}\right) = P(Z < -2.03) = \boxed{.0212} \text{ (LTT)}$$

This value can be found using TI-83: 2nd DIST → 2 (-∞, 11.61) enter.

Or by using table-1 for z as shown:

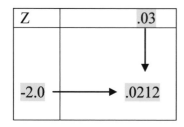

Z	.03
-2.0	.0212

.0212

If X is nearly normal binomial with mean $\mu = np$, and variance $\sigma^2 = npq$, then the standardized normal variance is:

$$Z = \frac{X - np}{\sqrt{npq}} \quad \dots (6)$$

Example-7

A sample of 100 data from binomial population with proportion P of success is 30%
Find the probability between 24 and 42.

Solution:

To find the probability means to find the area under the binomial curve between the
Two given points 24, and 42, the two point must be standard first by using the
Above formula (6);

$$P(24 < x < 42) = p\left(\frac{24 - np}{\sqrt{npq}} < Z < \frac{42 - np}{\sqrt{npq}}\right)$$

$$P\ (24 < x < 42\ ) = P\ (\dfrac{24 - (100)(30\%)}{\sqrt{100(3\%)\ (70\%)}} < Z < \dfrac{42 - (100)(30\%)}{\sqrt{100(30\%)\ (70\%)}})$$

$$= P\ (-1.3 < Z < 2.6) = \text{the area shown in the graph below:}$$

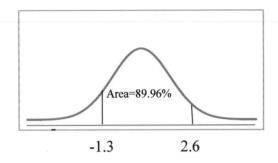

$$-1.3 \qquad 2.6$$

Example-8

For mean $\mu = 36$ and standard deviation $\sigma = 4.8$, find the following probabilities:

    a) $P(X \geq 45)$
    b) $P(x \leq 35)$

Solution:

To solve a – c first we have to standard the data, or change x to z using formula (6)

a) $P(X \geq 45) = P(Z \geq \dfrac{45 - 36}{4.8}) = p(Z \geq 2) = 2.28\%$

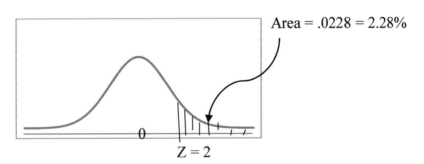

Area = .0228 = 2.28%

$$Z = 2$$

Using TI → {2nd DISTR (-∞, 2) enter} to get (-∞) → press {− 2nd , 99} enter

b) $P(x \leq 35) = P(Z \leq \dfrac{35 - 36}{4.8}) = P(Z \leq -0.21) = 48.68\%$

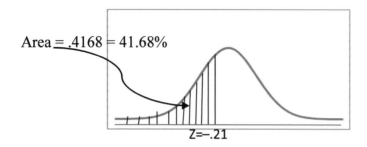

Area = .4168 = 41.68%

Z=−.21

5.3 Normal Distribution

Up to this section we've been dealing with distribution of **discrete variables**, but now in this section we will be dealing with probability distribution whose domain is the set of all real numbers or **continuous variables.** The model of this distribution is called the normal distribution, or "bell shaped", or the Gaussian distribution.

Normal distribution occurs in real applications, and they play an important role in methods of inferential statistics.

We will use the continuous variable (X) as the random variable with normal distribution that has a symmetric, and bell- shaped- curve. The measure of the probability is represented by the area under the bell-shaped-curve.

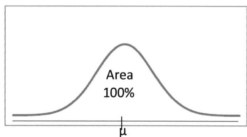

Area
100%

μ

The normal curve is also Unimodal and Symmetric about the mean.

As mentioned before, dealing with large number of data could be tedious, and the solution is To standard the data using formula called z-score:

$$Z = \frac{X - \text{mean of pop}}{\text{Standard deviation of pop}} = \frac{X - \mu}{\sigma} \quad \ldots \quad (6)$$

This z-score will give us indication of how far is the data from the mean.

If the normal model is labeled as: N (μ, σ) then the standard normal curve will represent N (0, 1), where $\mu = 0$, and $\sigma = 1$ for the standard normal curve $\{N(\mu, \sigma) \rightarrow N(0,1) \}$

$$Z = N(\mu, \sigma) \rightarrow N(0,1)$$

μ   0

The Empirical rule states that: 68% of the data falls within 1SD.
95% of the data falls within 2SD'S.
99.7% of the data falls within 3SD's.

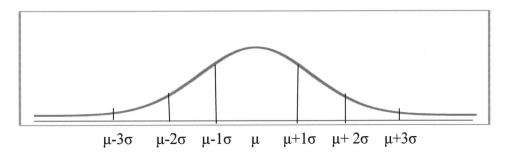

μ-3σ   μ-2σ   μ-1σ   μ   μ+1σ   μ+ 2σ   μ+3σ

Application of Normal Distribution:

Example-9

In a survey it was found that the weight of animals can be described by normal distribution model with standard deviation of 84 pounds and mean of 1152 pounds. What percentage of the animals weigh:
a) Over 1250 pounds.
b) Under 1200 pounds.
c) Between 1000 and 1100 pounds.

Solution:

a) $P(X > 1250) = p(Z > \dfrac{1250 - 1152}{84}) = P(Z > 1.67) = 0.0475$

Thus 4.75% of the animals weigh over 1250 ponds.

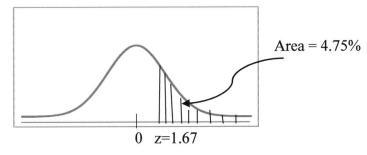

Area = 4.75%

0   z=1.67

Using TI: Area = 1- 2nd DISTR → 2 (-2nd 99, 1.67) enter

b) $P(X < 1200) = p (Z < \dfrac{1200 - 1152}{84}) = P (Z < .57) = 0.7157$

Or 71.57 of the animals weigh below 1200 pounds.

Area = 71.57

Z = .57

Using TI: Area = 2nd DISTR → 2 (- 2nd 99, 0.57) enter.

a) $P(1000 < X < 1100) = P(\dfrac{1000-1152}{84} < P < \dfrac{1100 - 1152}{84})$

$$= P (-1.81 < P < -0 .62) = 0.2325$$

Or 23.25% of the animals weigh between 1000 and 100 pounds.

Example-10

In a Stat-class, only the top 15% score A in the class. If the test has a mean of 200 with Standard deviation 20, find the lowest possible score to get A, assuming the scores are Normally distributed.

Solution:

In this question the area (probability) is given = 15%. So to find the score we have to Find Z.

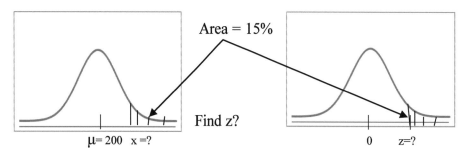

Area = 15%

Find z?

μ= 200   x =?

0   z=?

Find Z using TI→ 2nd DISTR → 3 (1-15%) enter → Answer: Z= 1.04

To find the score means to find the value of X, then change Z into X using the
Formula from (6):
X = Zσ + μ = (1.04)(20) + 200 = 220.8
Then students must score a minimum of 220.8 to get A score.

Determining Normality:

To determine when it is reasonable to assume that simple data are from population having a
normal distribution, we follow these steps:
- Construct a histogram for the data; if the data departs dramatically from the bell-shape, then we conclude that the data **do not** have a Normal Distribution.
- Identify the outliers, if there is more than one outlier present in the data, we conclude that the data **do not** have a normal distribution; sometimes even one outlier could cause an affect.
- Statisticians also check the skewness by using a formula called Pearson Coefficient given as:

$$PC = \frac{3(\bar{X} - Median)}{s} \quad \dots (7) \quad \text{Where, s=sample size}$$

Example-11

In a STAT-class the teacher gave her students a list of data and asked them to check
The normality of the data. Here is their solution:

Data: 62 52 43 71 5 50 29 13 56 66 55 58 17 69 59 30, n=16

Solution:

1. Construct the histogram use all the steps done in chapter-2:
   Let the number of classes be = 7.
   LV = 5, HV=71, R = 66, then W = 66/7 = 9.4 → round up to →10
   First class= 5 – 14.
   Construct the frequency distribution table:

Class Limits	Class Boundaries	Frequency F
5-14	4.4-14.5	2
15-24	14.5-24.5	1
25-34	24.5-34.5	2
35-44	34.5-44.5	1
45-54	44.5-54.5	2
55-64	54.5-64.5	5
65-74	64.5-74.5	3

We will use EXCEL to graph: See the steps on Chapter-2

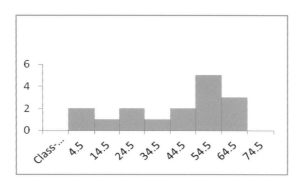

The Histogram shows departure from normality.

2. Checking the outlier(s):

We need the median and the quartiles, we use TI-83+ or TI-84+:

- STAT → EDIT → enter the data on L1.
- STST→ CALC → 1 enter→enter. The calculator will display:

$$
\begin{array}{c}
\text{Stat} \to \text{CALC} \to \text{1-Var} \\
X^- = 45.9 \\
S_x = 20.9 \\
LV = 5 \\
Q1 = 29.5 \\
Med = 53.5 \\
Q3 = 60.5 \\
HV = 71
\end{array}
$$

From the information taken from TI-83+ we find:

$$1.5(Q_3 - Q_1) = 46.5$$

$$Q_1 - 46.5 = -17 \qquad Q_3 + 46.5 = 107$$

Then the interval will be : $\{-17, 107\}$

This means all the data is falling inside of this interval, or there are no outliers in the list of data, but this does not prove abnormality.

3. Using the Pearson coefficient gives the significant skewness:

$$PC = \frac{3(X^- - Median)}{S} = \frac{3(45.9 - 53.5)}{20.9} = -1.1 \text{ Significant skewness.}$$

For Person Coefficient, if PC > +1 or PC <−1 → the data is considered significantly skewed.

Significant skewed -1   0   +1 Significant skewed

5.4 Sampling Distribution

In order to make inferences about the population, it is necessary to discuss the sample results further. The question one could ask: Is the sample mean $\overline{X}$ equal to the value of population mean μ? It is not! But we would hope it is at least close.

Sampling Distribution: The sampling distribution of a sample is the distribution of values for that sample statistics obtained from all possible Samples of a population. Samples must be of the same size.

Example-12

To illustrate the concept of sampling, consider the mean (μ) and range(R) of a sample of size 2 that can be drawn from the set of odd single digit integers {1, 3, 5, 7, 9}.

1. We can get 10-possible samples of size 2:

    (1,3) (3,5) (5,7) (7,9)
    (1,5) (3,7) (5,9)
    (1,7) (3,9)
    (1,9)

2. Each set has the following mean(μ):
    2   4   6   8
    3   5   7
    4   6
    5

3. Each of these samples is equally likely to occur with probability = 1/10
    Thus sampling distribution table and graph will be as shown:

$\overline{X}$	2	3	4	5	6	7	8
P(X)	1/10	1/10	2/10	2/10	2/10	1/10	1/10

And the probability distribution curve as shown below:

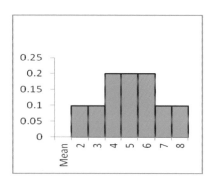

4. Now we will find the sampling for the range (R) of each set:
   Where R= HV– LV

$$
\begin{array}{cccc}
2 & 2 & 2 & 2 \\
4 & 4 & 4 & \\
6 & 6 & & \\
8 & & &
\end{array}
$$

Each value of R has equally likely probability of 1/10, and the distribution table with the graph is as shown below:

R	P(R)
2	.4
4	.3
6	.2
8	.1

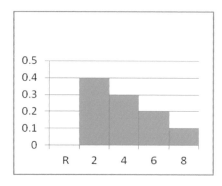

The Central Limit Theorem:

The central limit theorem concerns with the approximate normality of means of random samples of sum of random variables. Suppose: $x_1, x_2, \dots x_n$ is a sample from an infinite population with mean μ and standard deviation σ, where x, x, are independent random variables, then the central limit theorem gives:

1. The mean $\mu_{\bar{x}} = \mu$
2. The standard deviation is $\sigma_{\bar{x}} = \sigma / \sqrt{n}$
3. And the z-score is $z = (\bar{x} - \mu) / \sigma/\sqrt{n}$, and it is used to gain information about the mean.

Example-13

In a study about people's weight, a sample was collected, with mean 216, and standard deviation 23. If a person was selected at random,
  a) Find the probability that the person weighs less than 214.
  b) If a sample of 36 people was selected, find the probability that the mean of the sample will be less than 214.

Solution:

a) First we have to standardize the units: then,

$$P(X < 214) = P\left(Z < \frac{214 - 216}{23}\right) = P(Z < -.09) = .4641 = 46.41\%$$

Area = 46.41% | TI-83 → 2nd DISTR →2 (-∞,-.09)

z=-.07

b) $$P(\bar{x}\text{ (mean)} < 214)) = P\left(Z_{x-} < \frac{214 - 216}{23/\sqrt{36}}\right) = P(Z_{x-} < -.52) = .3015$$

Notice in this question the formula is different because a sample of n=36 Is involved.

Example-14

Find the z-value to the right of the mean so that 38.78% of the area under the distribution curve is in the left of it.

Solution:

Notice here the area is given and z is missing.

Z(area = 38.78%) → TI- 2nd DISTR →3 (normINV (38.78%) enter → – 0.29

Area given = 38.78%

Z= – o.29

Example-15

Find Z, if 72.10% of the area under the curve lies to the right of it.

Solution:

In the same as the above example using TI: 2nd DISTR → 3 (1-72.10%) = -.59
The area here is to the right, and tables and calculators normally read the area to the left, that is why we have to subtract it from 1.

Example-16

Consider a list of all possible samples of size 2 that are drawn from population with replacement: Let the set be {2, 4, and 6}.

1. First we will calculate the mean(μ) and standard deviation (σ) for the population using formulas:

Mean = $\mu = \Sigma$ X .P(X), and
Variance = $\sigma = \sqrt{\Sigma x^2. P(X)}$

X	P(X)	X.P(X)	X^2 . P(X)
2	1/3	2/3	4/3
4	1/3	4/3	16/3
6	1/3	6/3	36/3
		Σ=12/3=4	Σ=56/3

Mean = μ = 12/3 = 4, and
Standard deviation $\sigma = \sqrt{56/3 - 4^2} = 1.63$

1. Looking at all possible samples of size 2 from the population with replacement:

Possible samples	$\overline{X}$	$P(\overline{X})$	$\overline{X}\, P(\overline{X})$	$\overline{X}^2\, P(\overline{X})$
2,2	2	1/9	2/9	4/9
2,4	3	1/9	3/9	9/9
2,6	4	1/9	4/9	16/9
4,2	3	1/9	3/9	9/9
4,4	4	1/9	4/9	16/9
4,6	5	1/9	5/9	25/9
6,2	4	1/9	4/9	16/9
6,4	5	1/9	5/9	25/9
6,6	6	1/9	6/9	36/9
	Σ=36/9= 4		Sum = 36/9= 4	Sum = 156/9

From the table we see the mean $\mu_{x\text{-}}$ = 4
And the standard deviation $\sigma_{x\text{-}} = \sqrt{156/9 - 4^2} \approx 1.15$

5.5 Practice Problems

1. The probability that a randomly selected house has a garage space for 0, 1, or 2 cars is: 15%, 35%, 45% respectively. Construct a probability distribution for the data and graph the distribution.

2. A die is rolled in such a way that the probability of getting 1, 2,3,4,5, and 6 is: 1/2, 1/4, 1/6, 1/12, 1/12, 1/12 respectively. Construct a probability distribution and graph.

3. Find the following area under the normal distribution curve:
   a) Between z=0 and z=1.2.
   b) Between z=0, and z=-1.4.
   c) Between z=-1.14, and z=2.1.
   d) To the right of z=2.2.
   e) To the left of z=-.45

4. Find the z-value to the right of the mean so that:
   a) 66.7% of the area under the curve lies to the left of it.
   b) 67% of the area lies to the right of it.

5. Check the normality for the given data:
   25 54 64 45 55 66 38 47
   31 8   29 12 3   50 61 51

6 Confidence Intervals

Confidence Intervals

Statisticians collect data from experiments, observations or surveys and use them to draw conclusions about the phenomena used in the investigation, and they might use their data for the estimation of the values of unknown parameters or for tests of hypothesis concerning these values.

In this chapter we will be dealing with a sample from population involving unknown parameters; the problem is to construct a sample quantity that will serve to estimate the unknown parameter, such a sample quantity is called "estimator". A specific numerical value estimate of population is called a point estimate. The best point estimate of the population mean is the sample mean. From the measure of average we prefer to use the sample mean instead of the mode, or median, because the mean of the sample vary less than the other statistics. Sample measures are used to estimate population measures and are called estimators. Estimators should satisfy the following properties:

1. Estimator should be unbiased:
   For example: $E(s^2) = \sigma^2$ for all σ.

   $$E(\overline{X}) = \mu \text{ for all } \mu.$$

2. The estimator should be consistent as n-increases.

6.1 Confidence Intervals for the mean with known σ^2

A confidence interval (CI) is the interval of estimate of a parameter determined from sample data of confidence level estimate. The confidence intervals that we will deal with are: 90%, 95%, and 99%. The Central Limit Theorem states that, when sample size is large, then approximately 95% of the sample mean taken from population size will fall within ±1.96 standard errors of population mean i.e. $\mu \pm 1.96 \, (\sigma/\sqrt{n})$.

So the Central Limit Theorem states that: For large n, the mean will
Fall within → $\overline{X} \pm 1.96 \, (\sigma/\sqrt{n})$, or the interval is:

$$\overline{X} - 1.96 \, (\sigma/\sqrt{n}) < \mu < \overline{X} + 1.96 \, (\sigma/\sqrt{n})$$

And for specific α, the Central Limit Theorem when σ is known is:

$$\overline{X} - Z_{\alpha/2} (\sigma /\sqrt{n}) < \mu < \overline{X} + Z_{\alpha/2} (\sigma /\sqrt{n})$$
$$\overline{X} - Z_{\alpha/2} (\sigma /\sqrt{n}) < \mu < \overline{X} + Z_{\alpha/2} (\sigma /\sqrt{n})$$

CI	$Z_{\alpha/2}$
90%	1.65
95%	1.96
99%	2.58

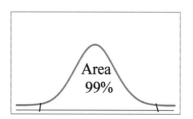

Required assumptions for CI when σ^2 is known
1. The sample is a random sample.
2. $n \geq 30$, if $n < 30$, then population must be normally distributed.

Example-1

Find the point estimate of population mean and the 90% CI of the population mean for the given data: Sample mean = 65, population standard deviation =7, with sample size = 60.

Solution:

The CI is → $\overline{X} - Z_{\alpha/2} (\sigma /\sqrt{n}) < \mu < \overline{X} + Z_{\alpha/2} (\sigma /\sqrt{n})$

$$65 - 1.96 \,(7/\sqrt{60}) < \mu < 65 + 1.96 \,(7 /\sqrt{60})$$
$$63.5 < \mu < 66.49$$

This means we are 90% confident that the mean fall between 63.5 and 66.49.

Check:

To check our work using TI follow these steps:
STAT → TESTS → 7 enter:

```
Input:     stat
σ : 7
x⁻ : 65
n : 60
C-Level: 90%
Calculate      enter
The calculator will display the following:
Z-interval (63.514, 66.486)
X⁻ = 65, n=60
```

Example-2

Find the 95% confidence Interval (CI) of the mean for the following data for a random of 30, assume σ=31.

Solution:

Data: 50 16 80 40 48 9   109 80 57 78 79 8   7   90 95
      6   44 4   52 4   64 21   49 33 15 3   54 29 8   47

Using TI → 1. STAT → EDIT → L1 (enter data)
           2. STAT → CALC → 1 enter, enter
           3. TI will display the following:

```
x⁻ = 42.633
Σ x = 1279
Σ x² = 83233
Sₓ = 31.46150722
σₓ = 30.9327047
n=30
```

Using the information from this table we can write the confidence interval:

$$CI \rightarrow \bar{X} - 1.96\,(\sigma/\sqrt{n}) < \mu < \bar{X} + 1.96\,(\sigma/\sqrt{n})$$

$$42.6 - 1.96\,(31/\sqrt{30}) < \mu < 42.6 + 1.96\,(31/\sqrt{30})$$

$$31.5 < \mu < 53.7$$

Or the interval = 42.6 ± 11.1. Hence we are 90% confident that the population
Of the data is between 31.5 and 53.7.
To check our work using TI →
1. STAT → EDIT → L1, enter the data on L1.
2. STAT → TESTS → 7 then add the values:

Input: Data	Z-interval (31.54, 53.726)
σ : 31	$\bar{x}$ = 42.635
C-Level: 90%	S_x = 31.46
Calculate enter	n =30
The calculator will display the following→	

6.2 Confidence Intervals for the mean with unknown σ^2

In this section we will find confidence interval for the mean with unknown variance, for a large
sample (n >30), or (n <30) with population normally distributed.
Since σ is unknown then it will be estimated by using sample variance (S). To keep the interval
at a given level we will use t-distribution. The t-distribution has characteristics some similar
others different from normal distribution as stated in the table below:

T-Similar Characteristics to ND	T- Different Characteristics from ND
1. The curve is bell-shaped.	1. Variance σ^2 >1.
2. The curve is symmetric about The mean.	2. T is a family of distribution with df=n-1
3. Measure of average falls at the Center.	3. Sample is random.
4. The curve never touches x-axis.	

The confidence interval about the mean for unknown variance is:

$$\overline{X} - t_{\alpha/2} (s / \sqrt{n}) \ < \ \mu < \ \overline{X} + t_{\alpha/2} (s / \sqrt{n}), \text{ with df} = n-1$$

The assumptions used for this case are:
1. Sample is random.
2. n ≥ 30, or n<30 with population normally distributed..

Example-3

Fifteen randomly selected students were asked: how much do they study for a test per week. The mean time was 5.4 hours, and the standard deviation 0.58. Find the 90% confidence interval of the mean time, assuming the variables are normally distributed.

Solution:

We have given: n=15, X =5.4, S=.58, CI=90%, and $\alpha = 10\%$, the interval is:

$$5.4 - t_{\alpha/2} (S /\sqrt{n}) \; < \; \mu < \; 5.4 + t_{\alpha/2} (S /\sqrt{n}),$$

We need to find $t_{\alpha/2}$ using table-2 as shown:
df = n-1 = 14

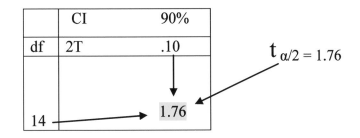

	CI	90%
df	2T	.10
14		1.76

$t_{\alpha/2} = 1.76$

Substituting, we get: $5.4 - 1.76 (.58 /\sqrt{15}) \; < \; \mu < \; 5.4 + 1.76 (.58 /\sqrt{15}),$

The CI-Intervals are: 5.14 ←——→ 5.66

To check our work using TI: STAT → TESTS → 8 (enter the information)

Input: Stats	T-interval (5.13622, 5.6638)
x⁻ : 5.4	x⁻ = 5.4
S_x :.58	S_x = .58
C-Level: 90%	n =15
Calculate enter	
The calculator will display the following→	

Example-4

A random sample was selected for the number of rainy days per year in a small town. Construct a 99% confidence interval for the data: 40   5   14   92   38   25   40   5   13   60

Solution:

We have n=10, df=9, CI=99%, and α =.01.
First we have to get the statistics from the data using TI:
1. STST $\rightarrow$ EDIT $\rightarrow$ L1 (enter data)
2. STAT $\rightarrow$ CALC $\rightarrow$ 1 ⊗the TI will display the following):

$$\bar{x} = 33.2$$
$$\Sigma x = 332$$
$$\Sigma x^2 = 17748$$
$$S_x = 27.3365851$$
$$\sigma_x = 25.93$$
$$n = 10$$

Then CI $\rightarrow$ 33.2 – t $_{\alpha/2}$ (s /$\sqrt{}$ n) < μ < 33.2 + t $_{\alpha/2}$ (s /$\sqrt{}$ n),
Now we use table-2 to find t $_{\alpha/2}$:

	CI	99%
df	2T	.01
9		3.250

$t_{\alpha/2}$

Then CI $\rightarrow$ 33.2 – 3.250 (27.34 /$\sqrt{}$ 10) < μ < 33.2 + 3.250 (27.34 /$\sqrt{}$ 10),
Or CI $\rightarrow$ 5.1 < μ < 61.3

Checking our work with TI:
STATS $\rightarrow$ TEST $\rightarrow$ 8 (then enter the information)
T-interval
Input: Data
$\bar{x}$: 5.4
List: L1
Freq:1
C-level: .99
Calculate: enter
The TI will display the following:$\rightarrow$

T-intervals
(5.1065, 61.293)
$\bar{x}$ = 33.2
S_x = 27.33658517
n=10

6.3 Confidence Intervals for Variance σ^2 and Standard Deviation σ

Since variance and standard deviation are as important as the mean, then we will find the confidence intervals (CI) for the variance and standard deviation too.
Here with CI for the variance and standard deviation a different statistical distribution is used called **"chi-square"** with symbol: χ^2, and a different table –G will be used to find the values of Chi-square, with degree of freedom given as df = n-1, and the table will give two values for the chi-square one to the left χ^2_L, and one to the right: χ^2_R

The chi-square variables are also a family of curves based on the degree of freedom df = n-2. The confidence interval for the variance is:

$$CI \rightarrow \frac{S^2(n-1)}{\chi^2} < \sigma^2 < \frac{S^2(n-1)}{\chi^2}$$

And the confidence interval for the standard deviation will be:

$$CI \rightarrow \sqrt{\frac{S^2(n-1)}{\chi^2_R}} < \sigma < \sqrt{\frac{S^2(n-1)}{\chi^2_L}}$$

The assumptions here are:
1. Sample is random.
2. Population must be normally distributed

Example-5

Find 90% confidence interval for the variance and standard deviation for a Sample with 15 and S=1.3.

Solution:

Here we have n=15, df =14, CI=90% $\rightarrow$ α = .10, and σ = 1.3.
First we have to find the two chi-square values using table-3 as shown:

$\alpha/2$=.05 $\alpha/2$=.05

Area
90%

df	CI	.90	df	CI	.90
		$\alpha = .95$			$\alpha = .05$
14		6.571	14		23.685
		χ^2_L			χ^2_R

Substituting the values of the right and left chi-squares, we get confidence intervals for the variance as:

$$CI \rightarrow \frac{(1.3)^2(15-1)}{23.685} < \sigma^2 < \frac{(1.3)^2(15-1)}{6.571} = 1 < \sigma^2 < 3.6$$

Then we conclude that we are 90% confident that the variance is between 1 and 3.6.
And confidence intervals for the standard deviation are:

$$1 < \sigma < 1.9$$

Thus we conclude that we can be 90% confident that the true variance is between 1 and 1.9.

Example-6

For a list of data: 67.7, 60.8, 62.4, 63.4, 75.8, 70.9, 72.3, and 73.3.
Estimate the population variance and standard deviation, for 95% CI.

Solution:

As done on the previous example, we use TI to get the missing information, and also use table-3 to find the two chi-squares as shown:

df	CI	.95	df	CI	.95
		$\alpha = .0975$			$\alpha = .025$
7		1.690	7		16.013
		χ^2_L			χ^2_R

After using TI to get the information that we need, we get:
For the variance: CI → $13.7 < \sigma^2 < 129.9$, and For the standard deviation
CI → $3.7 < \sigma < 11.4$.

6.4   Practice Problems:

1. In a center of 98 randomly selected adults showed that 87% of the respondents know what twitter is.
    a) Find the margin error (E) to a 95% confidence level.
    b) Find the 95% confidence interval.

2. How many statistics students must be randomly selected for IQ test, if we want 90% confidence that the sample mean is written 3IQ points of the population mean? (Note 3IQ means E = 3).

3. If the variance σ^2 =900, how large should a sample be to estimate the true mean score within 6 points with 95% confidence.

    Note: $E = (Z_{\alpha/2}\, \sigma) / \sqrt{n} \rightarrow n = (Z_{\alpha/2}\, \sigma) / E.$

7 Hypothesis Testing

Hypothesis Testing

$\mathcal{H}$ypothesis Testing is a decision making process for evaluating claims about a population. In performing hypothesis testing the following steps are applied:
- Define the population under the study.
- State the hypotheses that will be investigated.
- Give the significance level.
- Select a sample from the population.
- Collect the data.
- Perform the calculations required for the statistical test.
- Reach conclusion.

Parameter: is a numerical measurement that describes population.
Statistics: is a numerical measurement that describes sample

We have considered the basic idea of estimation for a parameter whose value was to be approximated based on a sample without a concern to the actual value of the parameter. We simply attempted to ascertain its value to the best of our ability. In contrast, when testing a hypothesis on a parameter, we have to be concerned about its value; this implies that 2-hypothesis is involved in any statistical study of this sort:

- The null hypothesis, denoted by H_0, and
- The alternative hypothesis, denoted by H_a.

The purpose of the experiment is to decide whether the evidence tends to reject the null hypothesis H_0. The following steps are used as guide line in hypothesis testing:

- Statement of equality always is included in H_0.
- We try to support H_a.
- We hope to reject H_0.

If H_0 is rejected → H_a is supported
If H_0 is not rejected → H_a is not supported

H_0 is rejected when test value falls inside the rejected region (RR)

Any time H_0 is rejected, Type-I error might be committed.
Any time H_a is rejected, Type-II error might be committed.

P (Type-I error) = α, P (Type-II error) = β

Errors of testing:

No matter how many evidence there are, it is still possible to make the wrong decision when performing a hypothesis test, we can make mistakes in two ways:
1. The null hypothesis H_0 is True, but we mistakenly reject it → Type-I error.
2. The null hypothesis H_0 is False, but we fail to reject it → Type-II error.

Q: Which one is more serious?
A: It depends on the situation under the test.

Example: In a jury-trial:

1. Type-I error occurs when the jury convicts an innocent person.
2. Type-II error occurs when the jury acquits a guilty person, and **this is more serious.**
This can be summarized by a tree-diagram:

Tree Diagram for the 4-cases of Hypothesis testing:

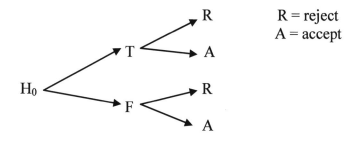

R = reject
A = accept

The 4-outcomes for the hypothesis testing are: {TR, TA, FR, FA},
Where:
TR = H_0 is true, but was rejected → Type-I error.
TA = H_0 is true, and was accepted → correct decision.
FR = H_0 is false, and was rejected → correct decision.
FA = H_0 is false, but was accepted → Type-II error.

We will perform hypotheses testing for the following parameters:
- Mean.
- Proportion.
- Variance and Standard Deviation.

The following table shows the type of test performed and the parameters with formulas used for each type.

Parameters	Type of Test	Formula used
Mean (μ)	z-test when σ is known	$z = \dfrac{\bar{x} - \mu}{\sigma/\sqrt{n}} \ldots (1)$
	t-test when σ is unknown $n \geq 30$, or $n < 30$ if pop is normally distributed	$t = \dfrac{\bar{x} - \mu}{s/\sqrt{n}} \ldots (2)$
Proportion (P)	z-test when np, nq ≥ 5	$z = \dfrac{\bar{P} - p}{\sqrt{pq}/\sqrt{n}} \ldots (3)$
Variance σ^2	χ^2-test	$\chi^2 = \dfrac{(n-1)\, s^2}{\sigma^2} \ldots (4)$

7.1   Hypothesis Testing for the Mean μ

Statistical Tests:
In this section two statistical tests will be used:
Z-test: This is used when σ (population standard deviation) is given.
T-test: This is used when S (sample standard deviation) is given.

In both cases n $\geq$ 30, if n <30 population must be normally distributed.

Z-Test

When applying the z-test we consider the followings:
- We find the critical value (CV) using TI, with given α.
- The test value (TV) is computed using formula (1).
- We assume the sample is random.
- The sample size n $\geq$ 30, or n < 30 with normally distributed data.

Solving Hypothesis-testing problems, the following steps are applied:
1. Write the 2-hypothesis H_0 (null hypothesis), and H_a (alternative hypothesis), and state which One is the claim. Then graph the normal curve and shade the required region.
2. Find the critical value(s) (CV) or the Z using TI or Z-table, and mark the value on the graph.
3. Compute the test value (TV) using the required formula, and mark the value on the curve.
4. Make the decision based on the two values from steps 3 and 4. If TV falls inside the shaded Region then reject H_0, otherwise do not reject H_0.
5. Summarize your decision.

Note: 2-types of problems will be used, one with given σ and x^- called "stat", and the other one with given data and we have to find σ and x^- called "data".

7.2   Z-Test for One-Mean

1. A. Z-Test for one-mean (μ) when σ is given (stat):

Example-1

An office manager claims that the average number of pages a person make using their copy machine is more than 30. A sample of 60 customer's orders was selected with mean of 31.1. At α = .05 is there enough evidence to support manager's claim? Use σ = 20.

Solution in steps:
1. The two hypothesis are:
   H_0: μ = 30
   H_a: μ > 30 (Claim), this is a Right-Tail-Test **(RTT)** as shown:

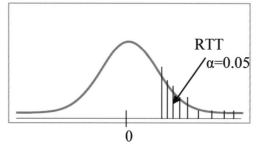

2. Find the critical value CV → or Z(α = 1– 0.05= 0.95) :
   Using TI: Press 2nd DISTR → 3(0.95) enter → answer ≈ 1.65.

3. Find the test-value (TV) using formula (1):

$$Z = \frac{\bar{x} - \mu}{(\sigma/\sqrt{n})} = \frac{31.1 - 30}{(20/\sqrt{60})} = 1.1/2.582 \approx 0.43$$

   Now on the graph mark both Z(CV)=1.65 and the test value (TV)= 0.43(red line) as
   shown below:

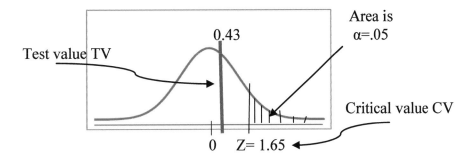

4. Since the z-test value (TV) the red-line is falling in the non-rejected region, we do not
   Reject the null hypothesis or H_0.
5. From this result we conclude that there was not enough evidence to support the claim H_a.

Check

To check our work using TI: we follow these steps:

1. STAT→TESTS→ 1(for z-test) enter the information:

   Z-Test
   Input: Stat   TI will display the following:
   μ_0: 30   Z-Test
   σ:20   μ > 30
   $\bar{x}$: 31.1   Z = .426... this agrees with
   n:60   our result.
   μ: >μ_0
   Calculate (enter)

2. To see if the graph match we follow all the above steps, but replace Calculate with →
   Draw (enter). We get the same graph as above, showing the test-value in the bottom of
   the graph.

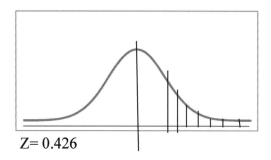

Z= 0.426

Example-2

The mean travel time, to work (in minutes) for a Co workers 16 years old and older, was 24.5 minutes. A random sample of 100 workers were selected and found that its mean was 22.5 minutes, and the population standard deviation is 5.35. At α = .05 can it be concluded that the mean time is less?

Solution in steps:

1. The two hypothesis are:
    H_0: μ = 24.5 minutes
    H_a: μ > 24.5 minutes (Claim), this is a Left-Tail-Test (**LTT**) as shown:

2. Find the critical value CV → or Z(α = 0.05) :
   Using TI: Press 2nd DISTR → 3(α = 0.05) enter → answer ≈ –1.65.

3. Find TV value using the formula in (1):

$$Z = \frac{\overline{X} - \mu}{(\sigma/\sqrt{n})} = \frac{22.5 - 24.5}{(5.35/\sqrt{100})} \approx -3.738$$

On the graph mark both Z (CV) = –1.65, and the test-value (TV) = – 3.74(red line) as shown below:

4. Since the test-value (red line) is falling inside the rejected-region, then we reject the null Hypothesis (H_0).

5. From this result we conclude that there was enough evidence to support the claim, which Is the alternative hypothesis H_a

Check

To check our work using TI: we follow these steps:

1. STAT→TESTS→ 1(for z-test) enter the information:

Z-Test	
Input: Stat	TI will display the following:
μ_0: 24.5	Z-Test
σ:5.35	$\mu < 24.5$
$\bar{x}$: 22.5	$Z = -3.73$ ← this (almost)agrees
n:100	with our result.
μ: $< \mu_0$	
Calculate (enter)	

2. To see if the graph match we follow all the above steps, but replace Calculate with → Draw (enter). We get the same graph as above, showing the test-value in the bottom of the graph.

Example-3

The average monthly rent for a one-bedroom home in a city is $1250. A random sample of 16 one-bedroom homes outside the city had a mean rent of $1350. The population standard deviation is $260. At α=0.05 can we conclude that the monthly rent outside the city differs from that in the city? Assume the population is normally distributed.

Solution in steps:

1. The two hypothesis are:

   H_0: $\mu = \$1250$

   H_a: $\mu \neq \$1250$ (Claim), this is a Two-Tail-Test (**2TT**) as shown:

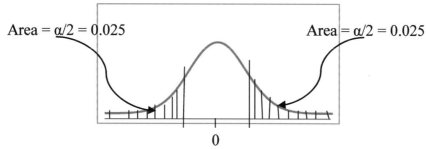

Area = α/2 = 0.025   Area = α/2 = 0.025

0

2. Find the critical value CV → or Z(α/2 = 0.025) :

   Using TI: Press 2nd DISTR → 3(0.025) enter → answer ≈ ± 1.96.

3. Find TV value using the formula (1):

$$Z = \frac{\overline{X} - \mu}{(\sigma/\sqrt{n})} = \frac{1350 - 1250}{(260/\sqrt{16})} \approx 1.54$$

On the graph mark both Z's (CV) = ± 1.96, and the test-value (TV) = 1.54(red line) as shown below:

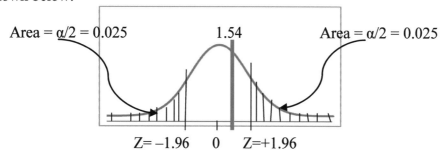

Area = α/2 = 0.025   1.54   Area = α/2 = 0.025

Z= –1.96   0   Z=+1.96

4. Since the test-value (red line), is falling outside the rejected-region we Do Not Reject the null hypothesis H_0.

5. From this result we conclude that there was not enough evidence to support the claim,
Which is, the alternative hypothesis H_a

| Check |

To check our work using TI: we follow these steps:

1. STAT→TESTS→ 1(for z-test) enter the information:

Z-Test
Input: Stat   TI will display the following:
μ_0: 1250   Z-Test
σ: 260 $\mu \neq 1250$
$\bar{x}$: 1350 $Z = 1.538$ ← this agrees with
n:16   our result.
$\mu: \neq \mu_0$
Calculate (enter)

2. To see if the graph match we follow all the above steps, but replace Calculate with →
Draw (enter). We get the same graph as above, showing the test-value in the bottom of
the graph.

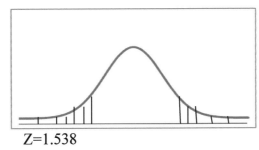

Z=1.538

| As mentioned earlier if data is given then we have to find all the missing
information from the given data, as shown in the next example |

1. B. Z-Test for one-mean (μ) with (data):

| Example-4 |

The protection agency keeps records of the miles/gallon of gasoline consumption
(in highways) for various makes and models of cars. For a random sample of 35
makes and models of cars, the collected miles/gallons data were:

30 25 33 23 49 20 29
27 24 24 20 52 25 27
22 15 30 18 24 27 24
25 35 25 27 24 24 28
24 35 24 10 18 32 21

Solution in steps:

Here, σ is not given we have to find it from the given data in the following steps:

a) Use TI to find the mean of the sample and standard deviations as follows:
   1. STAT → EDIT → L1 (enter the data in L1)
   2. STAT → CALC → 1 (1-variable) enter, enter TI will display the following:

1-Var Stats
$\bar{X} = 26.28871429$
$\Sigma X = 920$
$\Sigma X^2 = 26344$
$S_x = 7.9726..$
$\sigma_x = 7.8579...$
n=35

b) Conduct a hypothesis testing to see if the average number of mpg of gasoline is Less than 27 mpg, at 5% level of significance in the following steps:
   1. Write the hypothesis testing:
      H_0: μ=27
      H_a: $\mu < 27$ (claim) this is a Left-Tail –Test (LTT) as shown:

   2. Find the critical value (CV) or Z[α = 0.05]
      Using TI: press 2nd DISTR → 3(0.05) → answer ≈ – 1.65
   3. Find TV value using the formula (1):

$$Z = \frac{\bar{X} - \mu}{(\sigma/\sqrt{n})} = \frac{26.3 - 27}{(7.86/\sqrt{36})} \approx -.53$$

On the graph mark both Z (CV) = –1.65, and the test-value (TV) = – .53(red line) as Shown on the graph below:

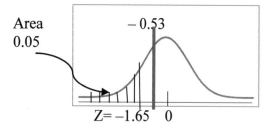

4. Since the test-value (red line), is falling outside the rejected-region we do not Reject the null (H_0) hypothesis.

5. From this result we conclude that there was not enough evidence to support The claim, which is the alternative hypothesis H_a.

Check

To check our work using TI: we follow these steps:

1. STAT→TESTS→ 1(for z-test) enter the information:

Z-Test	
Input: Data	TI will display the following:
$\mu_0 : 27$	Z-Test
S:7.97	$\mu < 27$
List:L1	$Z \approx -0.53$◄ this agrees with
Freq: 1	our result.
$\mu : < \mu_0$	
Calculate (enter)	

2. To see if the graph match we follow all the above steps, but replace Calculate with → Draw (enter). We get the same graph as above, showing the test-value in the bottom of the graph.

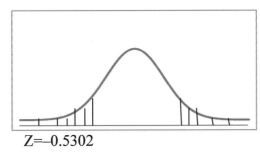

Z=−0.5302

Note: While finding the mean in all problems using TI, we can get The P-Value too which is given along with the z-value on TI And can be used for testing with the following rules:
If P-value $\leq \alpha$, Reject H_0.
If P-value $\geq \alpha$, Do not reject H_0.

Hypothesis Testing Between two Means μ_1 & μ_2:

When comparing the means of two populations, we typically consider the difference between their means, $\mu_1 - \mu_2$. The inference to be made about $\mu_1 - \mu_2$ will be based upon the difference between the observed sample means, $\bar{x_1} - \bar{x_2}$. This observed difference belongs to a sample distribution with the following characteristics:

> If independent samples of size n_1, n_2 are drawn randomly from a large populations with means μ_1 and μ_2, and variance σ_1^2 and σ_2^2, respectively, the difference between the means $\bar{x_1} - \bar{x_2}$, is approximately normally distributed with mean $\mu_1 - \mu_2$,
>
> And standard error $= \sqrt{\sigma_1^2/n_1 + \sigma_2^2/n_2}$

The first step in Hypothesis testing between two means (parameters), as was done in all Hypothesis testing is to state the 2-hypothesis testing (null H_0, and alternative H_a) and the claim, and the following 3-cases:

Left-Tail-Test (LTT)	Two-Tail-Test (2TT)	Right-Tail-Test (RTT)
$H_0 : \mu_1 = \mu_2$ OR $\mu_1 - \mu_2 = 0$	$H_0 : \mu_1 = \mu_2$ OR $\mu_1 - \mu_2 = 0$	$H_0 : \mu_1 = \mu_2$ OR $\mu_1 - \mu_2 = 0$
$H_a : \mu_1 < \mu_2$ OR $\mu_1 - \mu_2 < 0$	$H_a : \mu_1 < \mu_2$ OR $\mu_1 - \mu_2 \neq 0$	$H_a : \mu_1 < \mu_2$ OR $\mu_1 - \mu_2 > 0$

7.3   Z-Test for Two-Means μ_1 & μ_2

In the previous examples the Z-test for the mean was compared to a specific population mean. However, there are cases that require a comparison between two sample means. The following examples will test the difference between two means using Z-Test. The same 5-basic steps are used, as done in one-mean, with the following assumptions:
- Samples used are random samples.
- Samples must be independent of each other.
- The standard deviation of both populations must be given with sample size $n \leq 30$, and if $n < 30$, the population must be approximately or normally distributed. The Z- hypothesis testing formula for two means is:

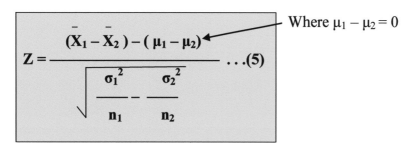

$$Z = \frac{(\bar{X_1} - \bar{X_2}) - (\mu_1 - \mu_2)}{\sqrt{\dfrac{\sigma_1^2}{n_1} - \dfrac{\sigma_2^2}{n_2}}} \quad \ldots(5)$$

Where $\mu_1 - \mu_2 = 0$

Example-5

A two independent samples of equal size of 40 were taken from two populations with equal standard deviation (0.6), and means $x_1^- = 2.01$, and $x_2^- = 2.19$. Using $\alpha = 0.05$, compare the two means.

Solution in Steps:

1. The hypothesis are:
   H_0: $\mu 1 = \mu 2$ or $\mu 1 - \mu 2 = 0$
   H_a: $\mu 1 < \mu 2$ **(LTT)**

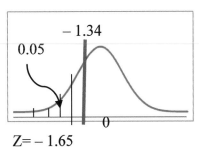

$Z = -1.65$

2. Find the critical value using Z-Test
   With-TI: we can find $Z(\alpha = .05)$ → 2nd DIST→ 3 (.05) → $z = -1.65$

3. Find the test value using formula (5):
$$Z = \frac{(2.01 - 2.19) - 0}{(\,(0.6)^2/40 + (0.6)^2/40)^{1/2}} = \frac{-0.18}{0.134} = -1.34$$

4. Since the test value falls inside the non-rejected region as shown in the graph above, then we **cannot reject** the null hypothesis (H_0), this means there is no difference between the two means.

Check

To check our work using TI: we follow these steps:

1. STAT→TESTS→ 3(for 2-sample z-test) enter the information:

Z-Test
Input: Stats   TI will display the following:
σ_1: 0.6   Z-Test
σ_2: 0.6 $\mu_1 < \mu_2$
x_1^-: 2.01 $Z = -1.34...$ ← this agrees with
x_2^-: 2.19   our result
n_1: 40
n_2: 40
μ: $< \mu_0$
Calculate (enter)

2. To see if the graph match we follow all the above steps, but replace Calculate with →
   Draw (enter). We get the same graph as above, showing the test-value in the bottom of
   the graph.

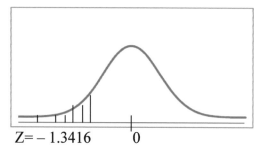

Z= − 1.3416   0

Example-6

Two independent samples are taken from two populations with the same variance of 900, and
with sample and mean as shown in the table below:

Sample	n	$\bar{x}$
A	30	35.6
B	35	38.3

Do these sample means provide sufficient evidence to reject the hypothesis that the means of
these two populations are equal? Use $\alpha = 0.05$.

Solution in Steps:

1. The hypothesis are:
   H_0: $\mu_A = \mu_B$ or $\mu_A - \mu_B = 0$ (claim)
   H_a: $\mu_A \neq \mu_B$ **(2TT)**

2. Find the critical value using Z-Test
   With-TI: we can find $Z(\alpha = .05)$ → 2nd DIST→ 3 (.05) → $z = \pm 1.96$

3. Find the test value using formula (5):
   $$Z = \frac{(35.6 - 38.3) - 0}{(900/30 + 900/35)^{1/2}} = -0.36$$

4. Since the test value (− 0.36) falls in the non-rejected region, as shown, then we do not
   Reject the null hypothesis or H_0.

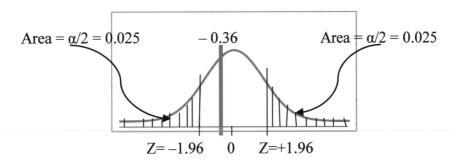

Area = α/2 = 0.025   – 0.36   Area = α/2 = 0.025

Z= –1.96   0   Z=+1.96

5. These sample means do not provide sufficient evidence to reject the hypothesis that the means
   Of these two populations are equal.

Check With TI

To check our work using TI: we follow these steps:
1. STAT→TESTS→ 3(for 2-sample z-test) enter the information:

Z-Test	
Input: Stats	TI will display the following:
σ_1 : 30	2-Sample Z-Test
σ_2: 30	$\mu_1 \neq \mu_2$
$\bar{x_1}$: 35.6	Z = – 0.3617◄— this is close to
$\bar{x_2}$: 38.3	our result
n_1: 30	
n_2 : 35	
μ: $\neq \mu_0$	
Calculate (enter)	

2. To see if the graph match we follow all the above steps, but replace Calculate with →
   Draw (enter). We get the same graph as above, showing the test-value in the bottom of
   the graph (Z= – 0.3617) which agrees with our results.

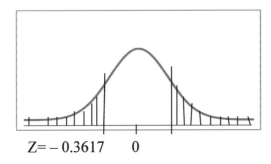

Z= – 0.3617   0

Example-7

Two sets of data are given as shown in the table. Test the claim that the two population means are different using α = 0.05.

Set A	23	24	18	29	21	21	26	27	18	19	23	21	28	26	19	22	18	22	$\sigma_A = 3.6$
Set B	18	22	22	19	19	19	22	28	22	19	36	23	25	21	18	27	20	24	$\sigma_B = 4.7$

Solution in Steps:

From the given data we use TI to find the mean of the sample and standard deviations as follows:

1. STAT → EDIT → L1 (enter the data in L1)
2. STAT → EDIT → L2 (enter the data in L2)
3. STAT → CALC → 1 (1-variable foe L1) enter, enter and,
4. STAT → CALC → 1 (1-variable foe L2) enter, enter and,

1-Var Stats for L1	1-Var Stats for L2
$\bar{X} = 22.5$	$\bar{X} = 22.44$
$S_x = 3.5355\ldots$	$S_x = 4.475$
n=18	n= 18

Now we can write the 5-steps for the hypothesis testing:

1. The hypothesis are:
   $H_0: \mu_A = \mu_B$ or $\mu_A - \mu_B = 0$
   $H_a: \mu_A \neq \mu_B$ **(2TT)** (claim)

2. Find the critical value (CV)using Z-Test
   With-TI: we can find Z(α = .05) → 2ⁿᵈ DIST→ 3 (.05) → z = ±1.96

3. Find the test value (TV) using formula (5):

$$Z = \frac{(22.5 - 22.44) - 0}{(3.6^2/18 + 4.7^2/18)^{1/2}} = 0.04$$

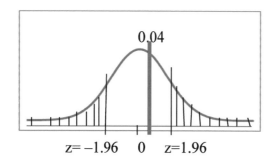

z= −1.96   0   z=1.96

4. Since the test value (– 0.36) falls in the non-rejected region, as shown, then we do not Reject the null hypothesis or H $_0$.

Check With TI

To check our work using TI: we follow these steps:
1. STAT→TESTS→ 3(for 2-sample z-test) enter the information:

Z-Test
Input: Stats   TI will display the following:
σ_1 : 3.6   2-Sample Z-Test
σ_2: 4.7 $\mu_1 \neq \mu_2$
$\bar{x_1}$: 22.5   Z = 0.0398..← this is close to
$\bar{x_2}$: 22.44   our result
n_1: 18
n_2 : 18
μ: $\neq \mu_0$
Calculate (enter)

2. To see if the graph match we follow all the above steps, but replace Calculate with → Draw (enter). The graph shows the test-value in the bottom of the graph (Z= 0.0398) which agrees with our results.

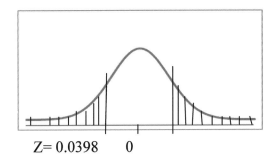

Z= 0.0398   0

Check using EXCEL

To check our work using EXCEL: we follow these steps:
1. Enter the first group of data in column-A on EXCEL with range A2:A16.
2. Enter the second group of data in column-B with range B2:B6.
3. Click on Data --> from Data Analysis --> select t-Test Two Sample Assuming Unequal Variances and click OK.
4. Enter the range for group-A as above.
5. Enter the range of the second group as above.
6. Enter 0 for hypothesized Mean Difference
7. Enter 0.05 for Alpha.
8. Click on output Range, and write D7.
9. Then click OK, EXCEL will display the following information:

z-Test: Two Sample for Means

	Variable 1	Variable 2
Mean	22.5	22.44444444
Known Variance	3.6	4.7
Observations	18	18
Hypothesized Mean Difference	0	
z	0.081813449	
P(Z<=z) one-tail	0.467397531	
z Critical one-tail	1.644853627	
P(Z<=z) two-tail	0.934795061	
z Critical two-tail	1.959963985	

T-Test

The T-distribution is similar to the standard normal distribution in many ways, but differs from it in curve wise. The T-distribution is a family of curves based on degree of freedom, where **degree of freedom=df=n–1**. Applying the statistical tests for the mean when S is given, and the population is normally or approximately normally distributed the T-test is used with the following formula that computes the test value for the mean:

$$t = \frac{\bar{x} - \mu}{s / \sqrt{n}} \quad \text{... (6) With df=n-1}$$

The same steps are used as in z-test, except that table-2 must be used for the critical value. Students need to practice on using table-2 to find the critical values; the following examples will help with that.

Example-8
Find the critical t-value for α=0.05 for a left tail test (LTT) with n=20.

Solution:

Since n=20, then the df = n–1 = 20–1 = 19. Then from table-2 find df=19 on the first column, and find α=0.05 on the first raw for one-tail, then read the value at the intersection, as shown in the table:

df	One tail α	0.25	0.10	0.05	0.025	0.01
	Two tail α	0.50	0.20	0.10	0.05	0.02
1						
2						
3						
.						
.						
19				1.729		

Then the critical value (CV) is (–1.729) for left tail.

Example-9

Find the critical t-value for α=0.01 for a right tail test (RTT) with n=26.

Solution:

Since n=26, then the df = n-1 = 26-1 = 25. Then from table-2 find df=25 on the first column, and find α=0.01 on the first raw for one-tail, then read the value at the intersection, as shown in the table:

df	One tail α	0.25	0.10	0.05	0.025	0.01
	Two tail α	0.50	0.20	0.10	0.05	0.02
1						
2						
3						
.						
.						
.						
25						2.787

CV

Then the critical value (CV) is (+2.787) for the right tail.

Example-10

Find the critical t-value for α=0.1 for a two tail test (2TT) with n=35.

Solution:

Since n=35, then the df = n-1 = 35-1 = 34. Then from table-2 find df=34 on the first column, and find α=0.1 on the first raw for one-tail, then read the value at the intersection, as shown in the table:

df	One tail α	0.25	0.10	0.05	0.025	0.01
	Two tail α	0.50	0.20	0.10	0.05	0.02
1				↑		
2						
3						
.						
.			CV			
.						
34			→	1.691		

The critical value is (± 1.691).
The test value can be computed by using formula (6).

7.4 T-Test for One-Mean μ

A. T-Test for one- mean (μ) when S is given (data):

Testing for the mean (μ) when σ is unknown → T-test is used, and the unknown σ is replace with the given sample standard deviation (S), or it can be found from the data (if data is given), the sample size must be greater than 30, or population normally distributed if sample seize is less than 30. Also, in T-test degree of freedom (df) is used where for a single mean is: df = n-1, and table-2 is used.

Example-11

A survey was conducted in a small town, about family sizes, the study reported that the average size of family in the town was 2.18, from the sample of 22 families that was collected as follows: (assuming the data is normally distributed)

$$3 \quad 4 \quad 3 \quad 3 \quad 5 \quad 2 \quad 3 \quad 3 \quad 4 \quad 3 \quad 4$$
$$2 \quad 2 \quad 3 \quad 3 \quad 4 \quad 3 \quad 6 \quad 1 \quad 2 \quad 2 \quad 5$$

Using α = .01, does the average family size differ from the claim.

Solution in steps:

1. The hypothesis are;
   H_0: μ = 2.18 (claim)
   H_a: μ ≠ 2.18, this is a **2TT**

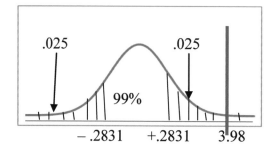

2. Find the critical value (CV) using table-2 with df=22-1=21 as shown:

df	2TT	.01
	CV	
21		.2831

3. Find the test-value using formula (2) or (6): the formula requires the mean of the sample $(\bar{X})$ we use TI: 2nd DISTR → EDIT → enter data in L1.
2nd DISTR → CALC → 1 (TI gives :)

1-Var stat	
$\bar{X} = 3.1818$	$S_x = 1.18065$
$\Sigma X = 70$	
$\Sigma X^2 = 252$	

Then using formula (6) we can find the t-value:

$$t = \frac{\bar{x} - \mu}{(s/\sqrt{n})} = \frac{3.18 - 2.18}{(1.18/\sqrt{22})} \approx 3.98$$

4. The test value 3.98 falls inside the rejected region RR (red line), which means we have to reject the null hypothesis H_0 (the claim). This means there was not enough information to support the claim.

Check

To check our work using TI: we follow these steps:
1. STAT→TESTS→ 2(for t-test) enter the information:

T-Test	TI-will display
Input (data)	T-Test
μ_0: 2.18	$\mu \neq 2.18$
List: L1	t=3.979← where this agrees with
Freq:1	$\bar{x} = 3.1818$ our result
μ: $\neq \mu_0$	$S_x = 1.180$
Calculate (enter)	

2. To see if the graph match we follow all the above steps, but replace Calculate with → Draw (enter). We get the same graph as above, showing the test-value in the bottom of the graph.

B. T-Test for one- mean (μ) when S is given (stat):

Example-12

According to the norms established for a test designed to measure a person's ability to think logically, college sophomore should average 82 points. This test is given to 41 randomly selected sophomores at a college. If they averaged 87 with a standard deviation of 10.2 on this test, can we conclude that the sophomore at that college think more logically than the average college sophomore? Use $\alpha = 0.01$.

Solution in steps:

1. The hypothesis are:
   H_0: $\mu = 82$ points
   H_a: $\mu > 82$ points (claim) this is a **RTT**

0   2.423

2. Find the critical value (CV) using table-2 with df=41-1=40 as shown:

df	1TT	0.01
		CV ▼
40	──────────►	**2.423**

3. Find the test-value using formula (2) or (6):

$$t = \frac{\bar{x} - \mu}{(s/\sqrt{n})} = \frac{87-82}{(10.2/\sqrt{41})} \approx 3.14$$

3.14

2.432

4. Since the test value 3.14 falls inside the rejected region, then we reject H_0.

5. We conclude that the sophomore at that college think more logically than the average College sophomore.

Check

To check our work using TI: we follow these steps:

1. STAT→TESTS→ 2(for t-test) enter the information:

T-Test	TI-will display
Input (stat)	T-Test
μ_0 : 82	$\mu > 82$
$\bar{x}$: 87	t=3.138.. ←where this agrees with
S_x : 10.2	our result
n : 41	$\bar{x} = 87$
μ: $> \mu_0$	n=41
Calculate (enter)	

2. To see if the graph match we follow all the above steps, but replace Calculate with →
   Draw (enter). We get the same graph as above, showing the test-value in the bottom of
   the graph t=3.1388

7.5 T-Test for Two-Means μ_1 & μ_2 | Independent Samples

The formula used to find the test-value (TV) for the T-test when σ is missing is:

$$t = \frac{(\bar{X}_1 - \bar{X}_2) - (\mu_1 - \mu_2)}{\sqrt{\dfrac{S_1^2}{n_1} - \dfrac{S_2^2}{n_2}}} \quad \dots(7)$$

Example-13

Upright vacuum cleaners have either a hard body type or a soft body type. Shown is the weight
(in ponds) of a sample of each type. At $\alpha = 0.05$ can it be conclude that the mean of the weights
are different?

Hard body type				Soft body type			
21	37	17	20	24	13	11	13
16	17	15	20	12	15		
23	16	17	17				
13	15	16	18				

Solution in steps:

From the given data we use TI to find the mean of the sample and standard deviations
As follows:

5. STAT → EDIT → L1 (enter the data in L1)
6. STAT → EDIT → L2 (enter the data in L2)
7. STAT → CALC → 1 (1-variable foe L1) enter, enter and,
8. STAT → CALC → 1 (1-variable foe L2) enter, enter and,

1-Var Stats for L1	1-Var Stats for L2
$\bar{X} = 18.625$	$\bar{X} = 14.67$
$S_x = 5.51210..$	$S_x = 4.76$
n=16	n= 6

Now we can write the 5-steps for the hypothesis testing:

1. The hypothesis are:
   H_0: $\mu_1 = \mu_2$ or $\mu_1 - \mu_2 = 0$
   H_a: $\mu_1 \neq \mu_2$ or $\mu 1 - \mu_2 \neq 0$ (claim) this is a two tail test **(2TT)**

2. To find the critical value we choose the smallest of the two degrees of freedom:
   $df_1 = n_1 - 1 = 16-1 = 15$
   $df_2 = n_2 - 1 = 6-1 = 5$
   Then the degree of freedom is = 5, using table-2 we can find the critical value CV:

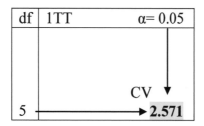

df	1TT	$\alpha = 0.05$
		CV ▼
5		→ **2.571**

3. Find the test value using formula (7):

$$t = \frac{(18.625 - 14.67) - 0}{\sqrt{(5.51)^2 / 16 + (4.76)^2 / 6}} \approx + 1.66$$

4. Since the test value (1.66) falls outside the rejected region, as shown, then we do not reject the null hypothesis or H_0.

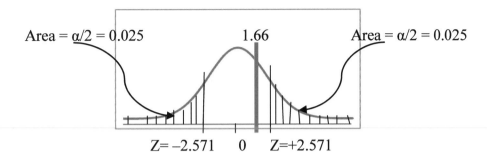

Area = α/2 = 0.025   1.66   Area = α/2 = 0.025

Z= –2.571   0   Z=+2.571

5. These sample means do not provide sufficient evidence to reject the hypothesis that the means of these two populations are different..

Check Using TI

To check our work using TI: we follow these steps:
1. STAT→TESTS→ 4(for 2-t-test) enter the information:

> 2 sample T-Test
> $\mu_1 \neq \mu_2$
> t=1.661..
> df = 20
> $x_1^- = 18.625$
> $x_2^- = 14.666...$
> $S_1 = 5.512..$
> $S_2 = 4.7609..$
> n1=16
> n2=6

2. To see if the graph match we follow all the above steps, but replace Calculate with → Draw (enter). We get the same graph as above, showing the test-value in the bottom of the graph as t=1.5501.

Check using EXCEL

To check our work using EXCEL: we follow these steps:
1. Enter the first group of data in column-A in EXCEL with range A2:A16.
2. Enter the second group of data in column-B with range B2:B6.
3. Click on Data --> from Data Analysis --> select t-Test Two Sample Assuming Unequal Variances and click OK.
4. Enter the range for group-A as above.
5. Enter the range of the second group as above.
6. Enter 0 for hypothesized Mean Difference
7. Enter 0.05 for Alpha.
8. Click on output Range, and write D7.
9. Then click OK, EXCEL will display the following information:

t-Test: Two-Sample Assuming Unequal Variances

	Variable 1	Variable 2
Mean	18.625	14.66667
Variance	30.38333	22.66667
Observations	16	6
Hypothesized Mean Difference	0	
df	10	
t Stat	1.661357	
P(T<=t) one-tail	0.063812	
t Critical one-tail	1.812461	
P(T<=t) two-tail	0.127625	
t Critical two-tail	2.228139	

The results agree with our calculation and TI.

7.6 T-Test for Two-Means μ_1 & μ_2 | Dependent Samples

Before, we used the t-test to compare two means for **independent** samples. Here we will use t-test to compare between two **dependent** samples. For dependent samples a different t-test will be used, and the 3 types of hypothesis testing LTT, 2TT, and RTT will be used as follows:

Left-Tail-Test (LTT)	Two-Tail-Test (2TT)	Right-Tail-Test (RTT)
$H_0 : \mu_d = 0$	$H_0 : \mu_d = 0$	$H_0 : \mu_d = 0$
$H_a : \mu_d < 0$	$H_a : \mu_d \neq 0$	$H_a : \mu_d > 0$

Where, μ_d is the mean of the dependent samples.

For simplicity we will be using the following symbols too:
- d = the difference between two data samples.
- μ_d = mean of the dependent samples.
- S_d = standard deviation of the differences.
- E= the error

The steps of the hypothesis testing are:
1. Find the difference between the data d = $(x_1 - x_2)$.
2. Find the mean of the differences $\mu = \sum d / n$
3. Find the standard deviation of the difference $S_d = \sqrt{\{n \sum d^2 - (\sum d)^2\} / n(n-1)}$
4. Find the standard error(estimated) $E = S_d / \sqrt{n}$
5. Find the test value (TV) using the following formula:

$$t = \frac{\mu - \mu_d}{S_d / \sqrt{n}} \quad \ldots \ldots \text{ (8) With df= n-1}$$

The assumptions for this type of test are:
- Samples are random.
- Data samples are dependent.
- When n< 30, the populations must be normally distributed.
- Normal curves are using as before.

Example-14

A sample of 10 grocery stores in a small town shows their profit (in millions of dollars) 3-years ago and their profits today. At $\alpha = 0.05$, can it be concluded that the average of their profits is more than that 3-years ago?

Sample	1	2	3	4	5	6	7	8	9	10
3-years ago	10.5	7.3	4.98	7.36	1.2	3.82	0.8	1.13	1.53	1.20
today	14.60	8.30	5.75	5.63	1.90	6.51	9.12	1.22	1.76	1.84

Solution in Steps:

Let the sample 3-years ago be X_1, and today be X_2. Then the above table will be:

X_1	10.5	7.3	4.98	7.36	1.2	3.82	0.8	1.13	1.53	1.20
X_2	14.60	8.30	5.75	5.63	1.90	6.51	9.12	1.22	1.76	1.84
$d = X_1 - X_2$	-4.1	-1	-.77	1.77	-.7	-2.69	-8.32	-.09	-.23	-.64
$d^2 = (X_1 - X_2)^2$	16.81	1	.5929	3.1329	.49	7.2361	69.2224	.0081	.0529	.4096

From the table we get: Mean of difference $\mu = \sum d/n = -16.77/10 = -1.677$
$\sum d^2 = 98.9549$, and $S_d = \{ 10(98.9549) - (-16.77)^2 / 10(9) \}^{1/2} = 2.8053\ldots$
Then the standard deviation of the difference is $S_d = 2.805$
Now the 5-steps of hypothesis testing are:
1. The hypothesis are:
   $H_0 : \mu_d = 0$
   $H_a : \mu_d < 0$ (claim) this is a Left Tail Test (LTT)

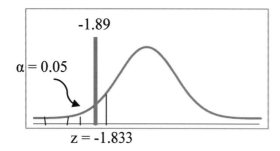

z = -1.833

2. Find the critical values for df=10-1 = 9 using the table as shown below:

df	1TT	0.05
		CV
9		▶ **1.833**

Then the critical value is – 1.833.

3. Compute the test value (TV) using the following formula:

$$t = \frac{\mu - \mu_d}{S_d / \sqrt{n}} = \frac{-1.677 - 0}{2.805 / \sqrt{10}} = -1.89$$

4. Since t-value (-1.89) falls inside the shaded region, then we reject H_0.
5. The claim can be supported for having enough evidence.

Check with TI

To check our work using TI: we follow these steps:

1. Enter the data from X_1 raw into column L_1, the data from X_2 into L_2 column.
2. On column L_3 replace it with (type) $L_1 - L_2$ then press enter, the TI will perform the difference on L_3.
3. Press STAT→TESTS→ 2(for t-test) enter the information:

T-Test	TI will display
Input: Data	T-Test
$\mu_0 = 0$	**t = –1.89853244**
List: L_3	$\overline{X} = -1.681$
Freq:1	$S_X = 2.79994623$
$\mu: < \mu_0$	n = 10
calculate (enter)	

4. To see if the graph match we follow all the above steps, but replace Calculate with →
Draw (enter). We get the same graph as above, showing the test-value in the bottom of
the graph as t= − 1.8985.

Check with EXCEL

1. Enter the two sets of 10-data each in one column of EXCEL sheet:

A	B
10.5	14.6
7.3	8.3
4.98	5.75
7.36	5.63
1.2	1.9
3.82	6.51
0.8	9.12
1.13	1.22
1.53	1.76
1.2	1.84

2. From Data Analysis tab, select t-test: Paired two Sample for Means.
3. Enter the Range for Column A as A2:A11.
4. Enter the Range for Column B as B2: B11.
5. Enter 0 for hypothesized Mean Difference
6. Enter Alpha = 0.05
7. Click on Output range and enter D5
8. Click OK. Then EXCEL will display the following information:

t-Test: Paired Two Sample for Means

	Variable 1	Variable 2
Mean	3.982	5.663
Variance	11.74188	18.13887
Observations	10	10
Pearson Correlation	0.755142	
Hypothesized Mean Difference	0	
df	9	
t Stat	**-1.89853**	
P(T<=t) one-tail	0.04505	
t Critical one-tail	1.833113	
P(T<=t) two-tail	0.090099	
t Critical two-tail	2.262157	

The information agrees with our calculation and TI.

Note: The same steps can be applied to problems of Right-Tail-Test, and Two-Tail-Test.

7.7 Hypothesis testing for Proportion

Many experiments require hypothesis testing for proportion. Proportion can be considered as Binomial experiment, with two-outcomes. We recall from Binomial distribution:
Mean (μ) = np, where n=sample size, and p=population proportion.
Population standard deviation (σ) = $\sqrt{npq}$.
Using Z-test for proportion:

- We use Normal Distribution to approximate the Binomial.
- We use formula (3) to find the test-value.
- Sample proportion $P^- = x/n$, where x is the data.

Assumptions for the testing:

- Sample is random.
- Conditions for Binomial must be met.
- $np \geq 5$, $nq \geq 5$.

Z-Test for Proportion

The formula used for the z-test for the proportion is:

$$Z = \frac{P^- - P}{\sqrt{Pq/n}} \quad \ldots\ldots\ldots (9)$$

Example-1

A statistician claims that 60% of students do not like to study on weekends. 200 students were selected randomly for the study and found that 108 stated that they do not like to study on weekends. At α =.05, is there enough evidence to reject the statistician's claim?

From the problems we gather the following information:
P = 60%
q=1– 60% = 40%
n=200, x=108

Solution in steps:

1. The hypothesis are:
   H_0: P = 60% (claim)
   H_0: P $\neq$ 60% (2TT)

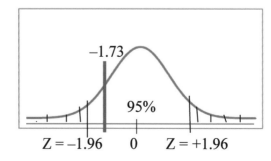

2. Find the critical value using TI: 2nd Dist → 3($\alpha = 0.025$) → Z= –1.96.
   Then the critical values are Z= ± 1.96

3. Find the testing value TV :
   $$P^- = x/n = 108 / 200 = 0.54$$

   Then using formula (8) gives: $Z = \dfrac{0.54 - 0.60}{\sqrt{\dfrac{(60\%)(40\%)}{200}}} \approx -1.73$

4. The test value falls inside the non-rejected region, thus **we do not reject H_0.**
   This means there is enough evidence to support the statistician's claim.

| Check |

To check our work using TI: we follow these steps:
   1. STAT→TESTS→ 5(for 1-proportion z-test) enter the information:

1-prop Z-Test	TI will display:
P_0 : .6	1-prop ZTest
x: 108	Z= –1.732..←This agrees with our results
n : 200	p = .083..
Prop ≠ p_0	p–= .54
Calculate (enter)	

   2. To see if the graph match we follow all the above steps, but replace Calculate with →
      Draw (enter). We get the same graph as above, showing the test-value in the bottom of
      the graph as Z= –1.7321.

| Example-2 |

   A survey on a group of children age 8-12 years, found that 17.4% were overweight.
   A random sample of 140 in this age group was selected for the study, and found that 20
   were considered overweight. At $\alpha = 0.05$ is this less that the national proportion?

From this problem we gather the following information:
P = 17.4%
q=1– 17.4% = 82.6%
n=140, x=20

| Solution in steps: |

   1.The hypothesis are:
   H_0: p = 17.4%
   H_a: p < 17.4% (claim) this is a right tail test **(LTT)**

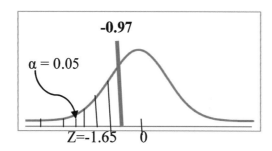

1. Find the critical value using TI: 2nd Dist → 3(0.05) → Z= –1.65
2. Find the testing value TV :

$$P^- = x/n = 20 / 140 = 0.143$$

Then using formula (8) gives: $Z = \dfrac{(20/140) - 17.4\%}{\sqrt{\dfrac{(17.4\%)(82.6\%)}{140}}} \approx -0.97$

3. The test value falls inside the non-rejected region, thus **we do not reject H₀.**
   This means there is not enough evidence to support the claim.

Check

To check our work using TI: we follow these steps:
1. STAT→TESTS→ 5(for 1-proportion z-test) enter the information:

1-prop Z-Test	TI will display:
P_0 : .174	1-prop Z-Test $p_0 < .174$
x :20	Z= –.97.... ←This agrees with our results
n :140	p = .1655....
Prop < p_0	p– = .1428..
Calculate (enter)	

2. To see if the graph match we follow all the above steps, but replace Calculate with →
   Draw (enter). We get the same graph as above, showing the test-value in the bottom of
   the graph as Z= –.972.

Example-3

The percentage of teachers in math department who are women is 30%. In a survey of
faculty of a large university, 45 of 130 randomly selected teachers were women. Is there
enough evidence at the 0.05 level of significance to conclude that the proportion of
women teachers at the university exceeds 30%?

From this problem we gather the following information:

P = 30%

q=1– 30% = 70%

n=130, x=45

Solution in steps:

1. The hypothesis are:

   H_0: p = 30%

   H_a: p > 30% (claim) this is a right tail test **(RTT)**

2. Find the critical value using TI: 2nd Dist → 3(0.05) → Z= +1.65
3. Find the testing value TV :

   $P^- = x/n = 45 / 130 = 0.35$

   Then using formula (8) gives: $Z = \dfrac{0.35 - 0.30}{\sqrt{\dfrac{(30\%\%)(70\%)}{130}}} \approx 1.24$

4. The test value falls inside the non-rejected region, thus **we do not reject H_0.** This means there is not enough evidence to support the claim.

Check

To check our work using TI: we follow these steps:

1. STAT→TESTS→ 5(for 1-proportion z-test) enter the information:

1-prop Z-Test	TI will display:
P_0 : .3	1-prop Z-Test p_0 >.3
x : 45	Z= 1.148...◄—This agrees with our results
n :130	p = .125......
Prop > p_0	p–= .346....
Calculate (enter)	

2. To see if the graph match we follow all the above steps, but replace Calculate with → Draw (enter). We get the same graph as above, showing the test-value in the bottom of the graph as Z= 1.1483.

7.8 Hypothesis testing for Variance and Standard Deviation

Hypothesis testing for Variance and standard deviation involves a different type of test called Chi-Square test with the symbol (χ^2), and use different table (3). We will use formula (4) to compute the test value.

Testing for a Single - Variance and standard deviation:

χ^2-Test for One-Variance & SD

Properties of Chi-Square distribution:

1. Chi-square is 0 or positive.
2. Chi-square in not symmetric, it is skewed to the right.
3. There is a different chi-square distribution for each degree of freedom.

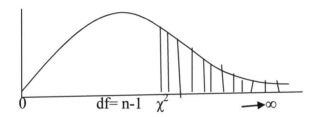

The formula to find the right Chi-square (χ^2_R) and left chi-square (χ^2_L) of the confidence interval of estimation for the variance are as shown in the table:

$$\chi^2_L = \frac{(n-1)\,S^2}{\chi^2\,(\,n-1,\,\alpha/2\,)} \quad , \quad \chi^2_R = \frac{(n-1)\,S^2}{\chi^2\,(\,n-1,\,1-\alpha/2\,)} \quad \dots (9)$$

As was done in the other tests, 3-cases are used in testing the variance (σ^2):

$H_0 : \sigma^2 = k$	$H_0 : \sigma^2 = k$	$H_0 : \sigma^2 = k$
$H_a : \sigma^2 < k$ for LTT	$H_a : \sigma^2 \neq k$ for 2TT	$H_a : \sigma^2 > k$ for RTT

Similarly, for the standard deviation (σ):

$H_0 : \sigma = k$	$H_0 : \sigma = k$	$H_0 : \sigma = k$
$H_a : \sigma < k$ for LTT	$H_a : \sigma \neq k$ for 2TT	$H_a : \sigma > k$ for RTT

To find the critical value (CV) for Chi-Square, table (3) is used and it is obtained based on degree of freedom df = n-1, and the area to the right of CV → in symbols we use:

For one-tail-test:
χ^2_R(df, α) to read the value of the area on the right.
χ^2_L(df, 1-α) to read the value for the area to the left.

For two-tail-test:
χ^2_R(df, α/2) to read the value of the area on right tail.
χ^2_L(df, 1-α/2) to read the value for the area on left tail.

To compute the **Test-value** the following formula is used:

$$\chi^2 = \frac{(n-1)\ S^2}{\sigma^2} \ \ldots\ldots\ldots\ (10)$$

Example-1
Find the critical value (CV) for chi-square with n=17, and α = 0.05.

Solution:

χ^2_R(df, α) = χ^2_R(n-1, α) = χ^2_R(16, 0.05) = 26.3 using the table as shown:

df	RTT	α = 0.05
		CV
16		26.3

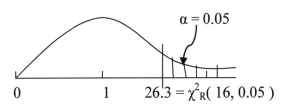

$26.3 = \chi^2_R$(16, 0.05)

Example-2
Find the critical value (CV) for chi-square with df=20, and $\alpha = 0.05$.

Solution:
$\chi^2_R(df, \alpha) = \chi^2_R(20, 0.05) = 31.4$ using the table as shown:

df	RTT	$\alpha = 0.05$
		CV
20		→ **31.4**

Example-3
Find the critical value (CV) for chi-square with df=14, and $\alpha = 0.90$.

Solution:

$\chi^2_R(df, \alpha) = \chi^2_R(14, 0.90) = 7.79$ using the table:

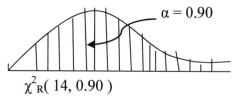

$$\chi^2_R(14, 0.90)$$

df	RTT	$\alpha = 0.90$
		CV
14		→ **7.79**

Solving problems of χ^2- distribution, the same 5-steps are applied as was done in the previous tests.

Example-4
For the given data (selected randomly):
  82   75   60   74   80   78   82   79   72   79 → n = 10
Is there sufficient evidence to conclude that the variance differs from 38? Use $\alpha=0.05$.

Solution in steps:
The formula requires both sample and population standard deviation. We use TI to Get them: STST→EDIT→ L1 (enter the data)

STAT → CALC → 1 (1-varable) enter, enter, TI will display:

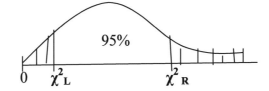

1-variable
$\bar{X} = 76.1$
$\Sigma X = 761$
$\Sigma X^2 = 58299$
$S_X = 6.55659$
$\sigma_X = 6.22$

1. The hypothesis testing are:
   $H_0 : \sigma^2 = 38$
   $H_a : \sigma^2 \neq 38$ (claim) (2TT)

2. Find the critical values using Chi-Square table-3 as follows: (df=10 –1=9)
   Note: Chi-square table reads area to the right. For two tails we read two critical values one on the right-tail χ^2_R (df, $\alpha/2$) = χ^2_R (9, 0.025) = 19.025, and on the left tail
   χ^2_L (df, 1-$\alpha/2$) = χ^2_R (9, 0.975) = 2.70

df	α	.975	.025
		↓	↓
9		2.700 →	19.025
		χ^2_L	χ^2_R

3. Find the test value using formula (10): With df=9, S≈6.6, and σ=40 from hypothesis testing = $9(6.6)^2/ 38 = \chi2 \approx 10.3$.

4. This value falls outside the rejected region, then the null hypothesis H_0 is not rejected, and the claim is not supported.

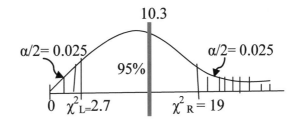

Example-5

In Stat-class of 28-students, the professor found in the mid-term exam of 100 points that the standard deviation σ=12. To study this he tested the above hypothesis at α=0.05 by giving the final exam to the same class of 28-students with standard deviation of 10.5.Does the professor have evidence that at level of α=0.05 the final test achieved lower standard deviation

Solution in steps:

1. $H_0 : \sigma = 12$
   $H_a : \sigma < 12$ (claim) this is a left tail test (LTT)

2. The critical value at α=0.05 for one-tail-test is:
   $\chi^2(df, 1-\alpha) = \chi^2_R (28-1, 0.95) = 16.151 \approx 16.2$

3. Compute the test-value using formula (10):

$$\chi^2 = \frac{(n-1)\,S^2}{\sigma^2} = \frac{(28-1)\,(10.5)^2}{(12)^2} = 20.671875 \approx 20.67$$

4. On the graph we see that the test-value (20.67) is more than the critical value (16.2) Or is not falling inside the rejected region, then we do not reject H_0, so we can not Support the professor's claim.
5. The professors needs more evidence to support his claim.

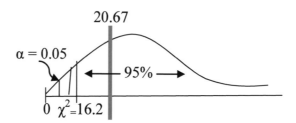

Example-6

A study shows that the weights of females have a standard deviation of no more than 10 pounds. Does the sample of 20 weights of female students in the first day date give sufficient evidence to reject the above claim? Use α=0.05 and S=11.9.

Solution in steps:

1. $H_0 : \sigma = 10$
   $H_a : \sigma > 10$ (claim) this is a left tail test (RTT)

2. The critical value at α=0.05 for one-tail-test is :
   $\chi^2(df, \alpha) = \chi^2_R (20-1, 0.95) = \chi^2 (19, 0.95) = 10.117$

3. Compute the test-value using formula (10):

$$\chi^2 = \frac{(n\text{-}1)\,S^2}{\sigma^2} = \frac{^{(}20\text{-}1)\,(11.9)^2}{(10)^2} = 26.9059 \approx 26.9$$

4. On the graph we see that the test-value (26.9) is more than the critical value (10.117) Or is falling inside the rejected region, then we do reject H_0, so we cannot support the Study.
5. There is enough evidence to support the claim.

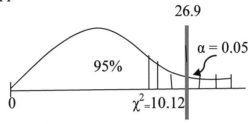

Testing for Two- Variances and Standard Deviations:

F-Test for Two-Variances

In research of different subjects different parameters are required for the study such as mean, proportion, variance and standard deviation. Here we will study the variance and standard deviation. Notice we are going to use different distribution called "F-test" with table (4) for the critical values.

> **Properties of F-distribution:**
> • F-distribution is positively skewed.
> • Mean value of F-distribution =1.
> • It is a family of curves depending on the degrees of freedoms df=n-1.
>
> The formula for F-test is:
>
> $$F = \frac{S^2_1}{S^2_2} = \frac{\text{Largest Variance}}{\text{Smallest Variance}} \dots\dots\dots (11) \text{ where } S^2_1 > S^2_2$$

F-test assumptions:
1. Samples are randomly selected.
2. Populations are normally distributed.
3. Samples are independent.

Note: Hypothesis testing steps for F-distributions are the same as was used in the other tests.

And the test-values, as was done in the other tests, 3-cases are used in testing the difference between two variances and standard deviations as shown:

$H_0 : \sigma_1^2 = \sigma_2^2$ $H_a : \sigma_1^2 \neq \sigma_2^2$ 2TT	$H_0 : \sigma_1^2 = \sigma_2^2$ $H_a : \sigma_1^2 > \sigma_2^2$ RTT	$H_0 : \sigma_1^2 = \sigma_2^2$ $H_a : \sigma_1^2 < \sigma_2^2$ LTT

Similarly, for the standard deviation (σ):

$H_0 : \sigma_1 = \sigma_2$ $H_a : \sigma_1 \neq \sigma_2$ 2TT	$H_0 : \sigma_1 = \sigma_2$ $H_a : \sigma_1 > \sigma_2$ RTT	$H_0 : \sigma_1 = \sigma_2$ $H_a : \sigma_1 < \sigma_2$ LTT

To find the critical value (CV) for F-test, table (4) is used and it is obtained based on degree of freedom df = n-1, and the area to the right of CV → in symbols we use:

For one-tail-test:
$F(df_n, df_d, \alpha)$ to read the value of numerator df_n, and denominator df_d.
$F(df_n, df_d, 1-\alpha)$

For two-tail-test:
$F(df_n, df_d, \alpha/2)$ to read the value of the area on right tail.
$F(df_n, df_d, 1-\alpha/2)$ to read the value for the area on left tail.

As we stated before, students should practice on finding the critical values using the required table, table (4) in here, before start solving real problems.

Example-7

Find the following critical values:
   a) F(24, 17, 0.05) Reading the table directly as shown gives (2.19):

	$\alpha = 0.05$
	df_n
df_d	24
.	
.	
.	
17	2.19

b)$F(5,8,0.05) = 3.69$

	$\alpha = 0.05$	
	df_n	
df_d		8
.		
.		
.		
5		→ 3.69

Example-8

A teacher wishes to find out if the variance of the grades achieved by students who study before the exam is different from those who do not study before the exam. He randomly selects two samples and found the following data:

Study before the exam	Do not study before the exam
$n_1 = 24$	$n_2 = 18$
$S_1^2 = 32$	$S_2^2 = 10$

Using $\alpha = 0.05$, is there enough evidence to support the claim?

Solution in steps:

1. $H_0 : \sigma_1^2 = \sigma_2^2$
   $H_a: \sigma_1^2 \neq \sigma_2^2$ (claim) this is a two-tail-test (2TT)

2. The critical value at $\alpha/2 = 0.025$ for two-tail-test from the 2-df's:
   $df_n = n_1 - 1 = 24 - 1 = 23$ (numerator degree of freedom)
   $df_d = n_2 - 1 = 18-1 = 17$ (denominator degree of freedom).
   Then table-(4) gives: $F(23,17,0.025) = 2.56$

	$\alpha = 0.025$	
	df_n	
df_d		24
.		
.		
.		
17		→ 2.56

3. Compute the test-value using formula (1):

$$F = \frac{S_1^{\,2}}{S_2^{\,2}} = \frac{32}{10} = 3.2$$

4. On the graph we see that the test-value (3.2) is larger than the critical value (2.56) Or is falling inside the rejected region, then we do reject H_0, so we cannot support the Study.

5. There is enough evidence to support the claim.

| **Check with TI** |

To check our work using TI: we follow these steps:
1. Press STAT→TESTS→ D (for 2-sampFtest) enter the information:

2-sampTtest	TI will display
Input: Stat	$\sigma_1 \neq \sigma_2$
S_1: 5.65685..	**F=3.2**
n_1: 24	
S_2: 3.16227...	
n_2: 18	
$\sigma_1: \neq \sigma_2$	
calculate (enter)	

1. To see if the graph match we follow all the above steps, but replace Calculate with → Draw (enter). We get the same graph as above, showing the test-value in the bottom of the graph as F=3.2.

F= 3.2

Example-9

A researcher wants to study the variance between the weight of third grade boys and girls (in pounds). Two samples were collected for the study as shown in the table. At $\alpha=.05$ can he conclude that the weight is different.

	n	S^2
Boys (B)	16	115.4
Girls (G)	25	138.5

Solution in steps:

1. $H_0 : \sigma_1^2 = \sigma_2^2$
   $H_a : \sigma_1^2 \neq \sigma_2^2$ (claim) this is a two-tail-test (2TT)

2. The critical value at $\alpha/2=0.025$ for two-tail-test from the 2-df's:
   $df_B = n_B - 1 = 16 - 1 = 15$ (denominator degree of freedom)
   $df_G = n_G - 1 = 25-1 = 24$ (numerator degree of freedom).
   Then table-(4) gives: F (df_G, df_B, $\alpha/2$) = F (24, 15, 0.025) = 2.70 as shown:

df_d	$\alpha = 0.025$
	df_n
	24
.	
.	
.	
15	2.70

3. Compute the test-value using formula (1):
$$F = \frac{S_1^2}{S_2^2} = \frac{138.5}{115.4} \approx 1.2$$

4. On the graph we see that the test-value (1.2) is less than the critical value (2.70)
   Or is falling outside the rejected region, then we do reject not H_0.

5. There is not enough evidence to support the claim.

Check with TI

To check our work using TI: we follow these steps:
1. Press STAT→TESTS→ D(for 2-sampFtest) enter the information:

2-sampTtest	TI will display
Input: Stat	$\sigma_1 \neq \sigma_2$
$S_{1:138.}$	
$n_{1:}$ 11.7686022	**F= 1.20017331**
S_2: 10.7424..	
List: L_3	
Freq:1	
$\sigma_1: \neq \sigma_2$	
calculate (enter)	

2. To see if the graph match we follow all the above steps, but replace Calculate with →
Draw (enter). We get the same graph as above, showing the test-value in the bottom of
the graph as F=1.2002.

F=1.2002

7.9 Practice Problems

1. People think taking one aspirin a day will decrease their blood pressure. If the mean of systolic blood pressure is 125, write the hypothesis and state the claim for the problem.

2. Using TI find the z for:
   a) $\alpha = .05$ for LTT.
   b) $\alpha = .10$ for RTT
   c) $\alpha = .04$ for 2TT.

3. A data for a sample of 20 is given as:
   20 27 30 29 23 24 25 29 31 27
   28 27 24 26 26 29 30 31 30 24
   For $\alpha = 0.10$ and population mean = 29:
   a) Find the sample mean.
   b) State the hypothesis claiming that the mean is not different from 29.
   c) Taking $\sigma = 2.92$ find the critical value using TI.
   d) Compute the test value.
   e) Draw conclusion, and explain your decision.
   f) Graph the curve showing all the values and the rejected region.

4. Using TI find the test-value for the following:
   a) n= 15, $\alpha= .005$ for LTT.
   b) n= 10, $\alpha= .05$ for RTT.
   c) n= 12, $\alpha= .10$ for 2TT.

5. For the given data:
   7 8 12 5 14 1 7 14 3 2
   a) Find the average of the sample, and the standard deviation S.
   b) State the hypothesis claiming that μ. 7.2
   c) Find the critical value using t-table using $\alpha = .05$
   d) Find the test value using taking $\mu = 7.2$.
   e) Write your decision and conclusion.
   f) Graph.

8 Correlation and Regression

Correlation and Regression

8.1 Correlation

Correlation is a statistical method to determine whether a linear relationship between two variables exists.

Regression is a statistical method to describe the nature of the relationship between two variables whether its positive, negative, linear, or nonlinear.

If the following questions are asked:

Q: Is there a relationship between height and weight?

Q: Is it true that the more you study, the better grades you achieve?

To answer these questions we need correlation.

The two variables involved are independent (X) and dependent (Y)

The types of relationship we get are:

- Simple relation → one independent variable.
- Multiple → two or more independent variables.
- Positive → both variables increase and decrease at the same time.
- Negative → As one variables increase the other one decrease and vice versa.

To study the relationship between two variables, we graph first so we can see the relation between the variables visually, and this is done by graphs called:

" **Scatter Plots".** Below is some type of graphs.

Scatter plot: is a plot of all the ordered pairs of bivariate data on a coordinate axes system. The input variable, x , will be used for the horizontal axis and the output variable ,y, for the vertical axis.

Note: When constructing a scatter diagram, we use scales along the x-axis so that the range of data along the vertical axis is equal or slightly shorter than the distance covered by the range along the horizontal axis.

A perfect linear correlation would arise where all the points fall exactly on a straight line, this could be either positive "when the line is sloping up or increasing", or negative when "the line is sloping down or negative". The numerical value of the calculated linear correlation coefficient "r" help us to find if there is a linear correlation between two-variables under study, and this value of **"r"** runs between ± 1, with the following assumptions:

a. If r is close to ± 1 → there is a linear correlation between the two variables.
b. If r is close to 0 → there is no correlation between the two variables.

Below is some type of graphs with different type of correlation as stated under each graph.

Positive Linear Correlation

No Correlation

High Positive Linear Correlation

Negative Linear Correlation

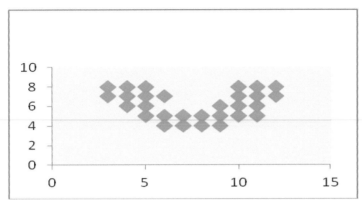

Non-Linear Relation

Scatter Plots:

The above graphs represent scatter plots which are graph of the ordered pairs (x, y) of numbers consisting of the independent variables (x), and the dependent variables (y).The scatter plots are a visual way to describe the nature of relationship between the variables under study.

Example-1

The following data show the number of hours (x) students study for test, and the grades they scored (y). Show the relationship between the two variables.

Hours of Study (X)	2	3	3	4	5	5	6	6	7	8
Grade (Y)	50	50	70	70	80	90	80	90	90	90

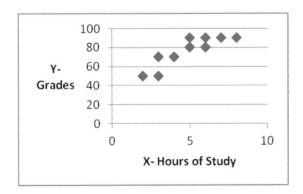

The Scatter-Plot for the data shows a positive
Relation between the two-variables x, and y.

Graph using TI

1. Press STST → EDIT → enter x- data on L1, and y- data on L2.
2. WINDOW: set up the window as follows
   Xmin = 0
   Xmax = 10
   Xscal = 2
   Ymin = 40
   Ymax = 100
   Yscal= 10
3. Press 2nd PLOT → enter → scatter → graph enter.

Graph using EXCEL

1. Enter X on cell A1, and Y on cell B1
2. Enter X-data on A2-A11, and Y-data on B2-B11.
3. Highlight both columns → insert → scatter → enter.

Correlation coefficient (r):

The correlation coefficient (r) measures the strength of the relationship between the Two variables.

Formula for r :

$$ r = \frac{n\,(\Sigma xy) - (\Sigma x)\,(\Sigma y)}{\sqrt{[n(\Sigma x^2) - (\Sigma x)^2][n(\Sigma y^2) - (\Sigma y)^2]}} \quad \dots (1) $$

Rules of r:
1. $-1 \le r \le +1$
2. If $r = -1$ there is a strong negative relationship.
3. If $r = +1$ There is a strong positive relationship.
4. If $r = 0$ There is no relation.

Computing the Coefficient (r):

Example-2

Use the data from example-1 to compute the coefficient r:

Solution in steps:

We will construct a table for all the required values then use the Formula (1) to compute(r):

X	Y	XY	X²	Y²
2	50	100	4	2500
3	50	150	9	2500
3	70	210	9	4900
4	70	280	16	4900
5	80	400	25	6400
5	90	450	25	8100
6	80	480	36	6400
6	90	540	36	8100
7	90	630	49	8100
8	90	720	64	8100
Σ =49	Σ = 760	Σ=3960	Σ=273	Σ=60000

To compute (r) we apply formula (1):

$$r = \frac{10(3960) - (49)(760)}{\sqrt{[10(273) - (49)^2][10(60000) - ((760)^2]}} = \frac{2360}{2714.70072} \approx 0.87$$

Based on the rules of correlation coefficient **r = 0.87** represents a strong positive Relation between(x) and (y).

Computing(r) using EXCEL

To check for (r) using EXCEL follow these steps:
1. Enter x-values on column-A-cells, and y-values on column-B-cells.
2. Select a blank cell → go to Formula (from tool bar) → click on f_x (insert function)
3. On the window of select category: select Statistical, and on select a function :Click on CORREL → OK, function argument window will open:
4. On CORREL widows enter: Array1: enter x- range
   Array2: enter y-range
5. Click OK and you get the answer = 0.869341 which agrees with our calculation above.

To see the relationship between x and y, graph using EXCEL:

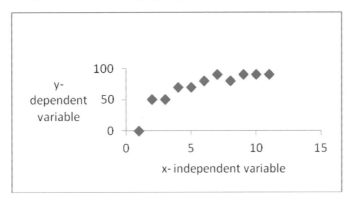

Note: r = sample correlation coefficient.
  ρ = Population correlation coefficient.
  df = n-2
  Testing is a 2-tail-test.
And for t-test we use the following formula:

$$t = r\sqrt{\dfrac{n-2}{1-r^2}} \quad \ldots (2)$$

Hypothesis testing for the population coefficient (ρ):

Using the same data from Example-1, we will try to test ρ.
From example-1 we have: r = 0.87, df = n-2 = 8, use α = .02

Solution in steps:

1. State the two hypothesis:
   H_0: ρ = 0 (means there is no population correlation)
   $H_{a:}$ $\rho \neq 0$ (means there is a population correlation)
   This is a 2-tail-test, as shown below:

.01 α =.01

98%

−2.896   +2.896

2. Find the critical values using table-2 as shown below:

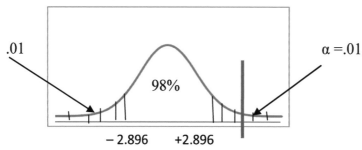

	CI	98%
df	2T	.02
8		**2.896**

3. Compute test-value using formula (2):

$$t = 0.87 \sqrt{\dfrac{8}{(.87)^2}} = 5.73657$$

4. Since the test-value **t = 5.74** (at the red line) is falling inside the critical region,
5. Then **Reject the null hypothesis**, this means there is a significant relation between the two variables.

Example-3

A study was conducted to predict the injuries cost by the fireworks from hundred million fireworks during a given year, the following data were obtained:

Fireworks (x)	Related Injuries (y)
67.6	12100
87.1	12600
117	12500
115	10900
118	7800
113	7000

    a. Draw a scatter plot for the data, and state the type of correlation from the graph.

    b. Compute the correlation coefficients.

    c. Perform the hypothesis testing in 5-steps, using α = 0.05.

Solution in steps:

a. We draw a scatter plot using EXCEL to find if there is a correlation between the 2-variables:

The graph shows a negative correlation.

a. To find the numerical value of this correlation we have to compute r , but first we have to complete the above table, then compute r using the following formula:

$$r = \frac{n(\sum xy) - (\sum x)(\sum y)}{\sqrt{[n(\sum x^2) - (\sum x)^2][n(\sum y^2) - (\sum y)^2]}}$$

Firework (x)	Related Injuries y)	X²	Y²	xy
67.6	12100	4569.76	146410000	817960
7.1	12600	7586.41	158760000	1097460
117	12500	13689	156250000	1462500
115	10900	13225	118810000	1253500
118	7800	13924	60840000	920400
113	7000	12769	49000000	791000
617.7	62900	65763.17	690070000	6342820

Substituting the values from the table into r-formula above gives:

$$r = \frac{6(6342820) - (617.7)(62900)}{\sqrt{[6(65763.17) - (617.7)^2][6(690070000) - (62900)^2]}}$$

$$r = \frac{38056920 - 38853330}{\sqrt{(13025.73)(184010000)}} = \frac{-796410}{1548181.054} \approx -0.514$$

b. As indicated in part (a) the correlation is negatively skewed. To find how strong is the correlation or how significant, we have to perform the 5-steps of hypotheses testing using a parameter called rho (ρ) represents the sample correlation coefficient for all the possible pairs.

1. $H_0: \rho = 0$
   $H_a: \rho \neq 0$ (claim) (2TT)

2. Find the critical value with the α=0.05, and df=n-2=4, using table or TI as:
   t(df, α) = t(4, 0.05 2TT) = 2.776

df	2TT	α=0.05
.		
.		
.		
.		
4		±2.776

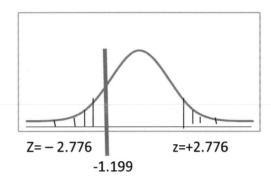

Z= − 2.776   z=+2.776

-1.199

3. Compute the test-value using the formula: $t = r \sqrt{\frac{n-2}{1-r^2}} = -0.514 \sqrt{\frac{4}{1-(-0.514)^2}} \approx -1.199$

4. Since the test value is falling in the non-rejected region, and then we do not reject H_0.

5. We conclude there is not sufficient evidence to support the claim.

A Geometric way of Measuring (r):

In Geometry we deal with conic shapes such as circle, ellipse, parabola, hyperbola,.., etc.
If we frame our scatter plot that has a shape of ellipse with an elliptical border then geometrically we can measure (r) as follows:

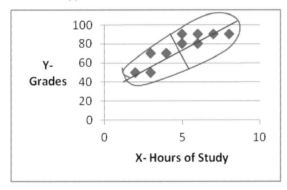

a. Measure the long diameter = D.
b. Measure the short diameter = d.
c. Compute r using the following formula: $r = \pm (1 - d/D)$ where (+) is used for a positive scatter-plot (increasing), and (−) indicates a negative scatter plot (decreasing).

The strength of the correlation:

To measure the strength of the correlation between (x), and (y), we perform regression analysis, first by finding the **line of best fit** and its equation, using the information given in the problem. Since the line of best fit is just a straight line, then the equation of the line of best fit is simply the equation of straight line: $y = \pm mx + b$, (+) for positive or increasing line, and (−) for negative or decreasing line. Then we need to find m (the slope) and b (the y-intercept of the line), these are given by the following formulas:

Slope of the line is: m = $\dfrac{n(\sum xy)-(\sum x)(\sum y)}{[n(\sum x^2)-(\sum x)^2]}$

The y-intercept of the line is: b = $\dfrac{n(\sum y)(\sum x^2)-(\sum x)(\sum xy)}{[n(\sum x^2)-(\sum x)^2]}$

These steps are done in analysis of linear regression in the next section.

8.2 Linear Regression

One of the primarily objective in regression analysis is making predictions, i.e. the prediction that student success is based on the hours of study, or predicting the distance required to stop a car is based on its speed. So the relationship between the dependent and independent variables depends on the **prediction equation**, whether is a straight line equation, a quadratic equation or nonlinear equation, as described graphically:

1. We predict a linear relationship when the equation is of the form:
   y = a x + b (or equation of a straight line)

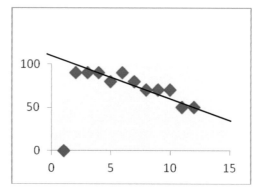

Linear regression with positive slope   Linear Regression with negative Slope

2. We predict a quadratic relation when the equation is:
    $y = a x^2 + b x + c$ (or the quadratic equation)

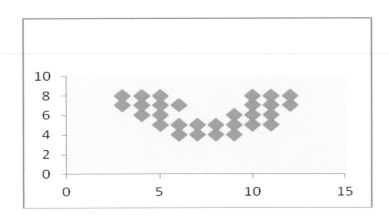

Non-Linear or Quadratic equation

3. We predict an exponential or logarithmic relation when the equation is:

$y = a b^x$ (or no relation)

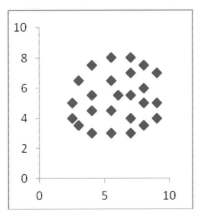

If the relationship is of a straight line, then the best fitting straight line is found by using the method of least square.

Example-1

A study was conducted in a school district to predict the number of teachers who voluntary resign, the following data were obtained from the school district record:

Number of years in service (x)	Number of teachers who voluntary resigned (y)
15	10
9	
13	16
11	14
12	15
	15

Do a complete regression analysis by performing the following steps:

a. Draw a scatter plot for the data, and state the type of correlation from the graph.

b. Compute the correlation coefficients.

c. Perform the hypothesis testing in 5-steps, using $\alpha = 0.05$.

d. Determine the regression line equation.

e. Find the number of teachers that would resign among those who have had 14 years Of service.

f. Plot the regression line on the scatter plot.

g. Summarize your results.

Solution in steps:

a. We draw a scatter plot using EXCEL to find if there is a correlation between the 2-variables:

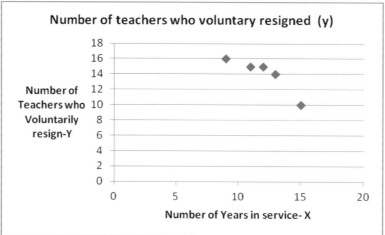

The graph shows a high negative correlation.

b. To find the numerical value of this correlation we have to compute r , first we have to complete the above table, then compute r using the following formula:

$$r = \frac{n(\sum xy) - (\sum x)(\sum y)}{\sqrt{[n(\sum x^2) - (\sum x)^2][n(\sum y^2) - (\sum y)^2]}}$$

(x)	(y)	x^2	y^2	xy
15	10	225	100	150
9	16	81	256	144
13	14	169	196	182
11	15	121	225	165
12	15	144	225	180
60	70	740	1002	821

Substituting the values from the table above into r-formula gives:

$$r = \frac{5(821) - (60)(70)}{\sqrt{[5(740) - (60)^2][5(1002) - (70)^2]}} = \frac{4105 - 4200}{\sqrt{(100)(110)}} = -0.88$$

c. As indicated in part (a) the correlation is highly negative. To find how strong is the correlation or how significant, we have to perform the 5-steps of hypotheses testing using a parameter called rho (ρ) represents the sample correlation coefficient for all the possible pairs.

1. $H_0: \rho = 0$ (claim)
   $H_a: \rho \neq 0$ (2TT)

2. Find the critical value with α=0.05, and df=n-2=6-2 = 4, using table or TI as:
   t(df, α) = t(4, 0.05 2TT) = 2.776

df	2TT	α=0.05
.		
.		
.		
.		
4		±2.776

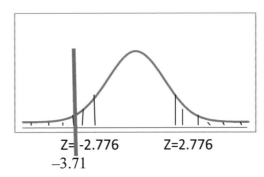

Z= -2.776   Z=2.776
−3.71

3. Compute the test-value using the formula: $t = r \sqrt{\frac{n-2}{1-r^2}} \approx -3.71$

4. Since the test value is falling in the rejected region, and then we reject H_0.

5. We conclude there is not enough evidence to support the claim.

d. To measure the strength of the correlation between (x), and (y), we perform regression analysis, first by finding the **line of best fit** and its equation, using the information given in the above table: Since the line of best fit is just a straight line, then the equation of the line of best fit is simply the equation of straight line: $y = \pm mx + b$, (+) for positive or increasing line, and (–) for negative or decreasing line. In our case the equation is increasing or : $y = mx + b$. Then we need to find m (slope) and b (the y-intercept of the line), these are given by the following formulas:

$$m = \frac{n(\sum xy) - (\sum x)(\sum y)}{[n(\sum x^2) - (\sum x)^2]} \approx -0.95$$

$$b = \frac{(\sum y)(\sum x^2) - (\sum x)(\sum xy)}{[n(\sum x^2) - (\sum x)^2]} = 25.4$$

Hence the equation of regression line with negative slope is: **Y= – 0.95 X + 25.4**

To draw the line of best fit, we choose two values for x (9, 11) substitute them in the regression line to find 2-values for y:

Y1 = - 0.95(9) + 25.4 = 16.85 → the point is: (9, 16.85)
Y2 = -0.95(11) + 25.4 = 14.95 → the point is: (11, 14.95)

Mark down these 2-points on the scatter plot, and connect between them to get the line of best fit. To find the mid-point $(\bar{X}, \bar{y}) = (\frac{\sum x}{n}, \frac{\sum y}{n}) = (12, 14)$.

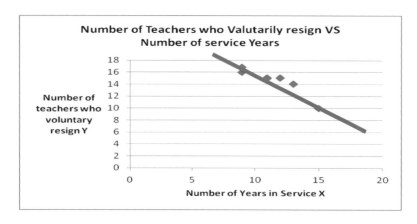

Example-2

To find the relationship between height (in inches) and weight (in pounds) of women, data were collected from 8-women as shown on the table:

Height (X)	64	64	61	66	68	64	60	66
Weight (Y)	104	124	109	119	139	134	94	129

Solution in steps:

1. Draw a scatter plot to see if there is a relation between the two variables.
   Using TI-83+ or 84+: Follow these steps:
   STAT → EDIT → enter x-values on L1, and y-values on L2.
   Set up window: Xmin = 60, Xmax= 70, Xscal= 2
                        Ymin = 90, Ymax = 130, Yscal = 10.
   2nd Stat-Plot → L1, L2 → Graph

Using EXCEL to graph:

1. Enter x-data in A, y-data in B.
2. Highlight the two columns and press scatter.

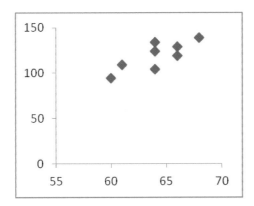

The graph shows a linear and positive correlation.

To compute the numerical value of (r), we have to complete the table with the required Values as shown in step-2:

2. Calculate the linear correlation coefficient **(r)** using formula (1)
To do this we need to construct a table of all the required items for the formula:

X	Y	XY	X^2	Y^2
64	104	6656	4096	10816
64	124	7936	4096	15376
61	109	6649	3721	11881
66	119	7854	4356	14161
68	139	9452	4624	19321
64	134	8576	4096	17956
60	94	5640	3600	8836
66	129	8514	4356	16641
$\Sigma = 513$	$\Sigma = 952$	$\Sigma = 61277$	$\Sigma = 32945$	$\Sigma = 114988$

Using these information apply formula (1) to compute **r ≈ 0.98**

1. Calculate the equation of the best fit:

The estimated equation is: $\overline{Y} = m\overline{X} + b \rightarrow b = \overline{Y} - m\overline{X}$

$\overline{Y} = \Sigma Y / n = 952/8 = 119$

$\overline{X} = \Sigma X / n = 513 / 8 = 64.13$

$$ m = \frac{n(\Sigma xy) - (\Sigma x)(\Sigma y)}{[n(\Sigma x^2) - (\Sigma x)^2]} = \frac{8(61277) - (513)(952)}{8(32945) - (513)^2} = \frac{1840}{391} = 4.71 $$

$b = 119 - (4.71)(64.13) = -183.05$

Then the estimated $Y = -183.05 + 4.71 X \dots (1)$

2. Draw the line of best fit by choosing 2- x points and substitute in equation (1)
to get y:
For $X_{min} = 60 \rightarrow$ Y(estimated) $= -183.05 + 4.71(60) = 99.55 \rightarrow (60, 99.55)$
For $X_{max} = 66 \rightarrow$ Y(estimated) $= -183.05 + 4.71(66) = 127.81 \rightarrow (70, 127.81)$
Mark the two points (60,99.55) and (70, 127, 81) on the graph and connect between
them to get the **line of best fit** as shown: The mid point is $(x^- = \Sigma x/n, y^- = \Sigma y/n)$
Mid Point= (513/8, 952/8) = (64.13, 119)

These two points will give the **line of best fit** as shown below in the first graph, and on the second graph the line is drawn on the scatter plot.

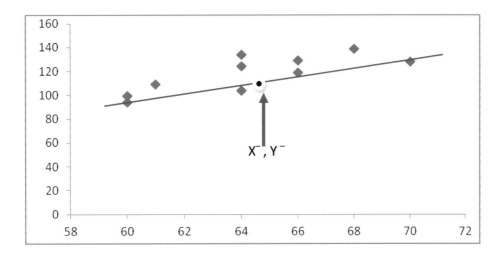

The above graph shows the data, and the two points, with the line of best fit Connected between them, also showing the point (x̄, ȳ) falling on the line.

8.3 Practice Problems

1. A marketing firm wishes to determine whether or not the number of refrigerator Commercials broadcast were linearly correlated to the sales of its product. The data obtained from each of several cities were as follow:

Refrigerator (x)	10	7	8	14	10	14	7	15
Sales(y)	7	4	9	13	11	8	6	10

   a) Draw a scatter plot, and estimate (r).
   b) Calculate (r).
   c) Is there evidence that the number of commercials is linearly correlated to the sales?

2. Teachers believe that students 'grades are affected by their attendance, to test The claim, a teacher collected data from her record of fall grades:

No of absence (x)	8	10	1	0	6	4
Grade (y)	68	62	94	92	75	80

   a) Draw a scatter plot.
   b) Compute the correlation coefficient.
   c) State the hypothesis.
   d) Test the hypothesis at $\alpha=.10$
   e) Determine the regression linear equation.
   f) Plot the regression line on the scatter plot.
   g) Summarize the results.

9 Analysis of Variance

9.1 Constructing ANOVA table
9.2 Practice Problem

Analysis of Variance
ANOVA

$\mathcal{T}$he analysis of variance is an arithmetic device for partitioning the total variation in a set of data according to the various sources of variation that are present. It results in a table of analysis of variance for simple linear regression; this table provides summary for the information contained in a set of data. Study of the complete analysis – of – variance table for a set of experimental or sample survey data will show whether or not valid tests of certain hypothesis exist, and if so, how the test should be performed.

The Model

New symbols will be used in this chapter as follow:
X_{ij} represent an observation with subscript (i) for the sample, and (j) for the individual observation within sample, for example: X_{31} is the 3rd sample with 1st observation.
The mathematical model is:

$$X_{ij} = \mu + \tau_i + \varepsilon_{ij} \dots (1)$$
Where, i = 1, 2 ... k, and j= 1, 2, . . . n

Where,
μ = overall mean
τ_i = the derivation of the i-th population
ε_{ij} = Random deviation from the mean of the i-th population.
For example: If μ_i = the mean of the ith population → then $\mu = \Sigma \ \mu_{i/k}$
$\tau_{i = \mu_i - \mu}$, and $\varepsilon_{ij} = x_{ij} - \mu_i = x_{ij} - \mu - \tau_i$
For this model we will make the following assumptions:
      1. μ is unknown.
      2. τ_i is unknown.
      3. ε_{ij} are normally and independently with mean =0, and variance $\sigma^{2.}$

9.1 Constructing ANOVA table

ANOVA table is represented by:

Source	df	Sum of squares	Mean squares
Between groups	k-1	SS_B	MS_B
Within groups	N-K	SS_W	MS_W
Total			

Where,

df=degree of sum

SS_b = sum of squares between the groups.

SS_w= Sum of squares within the groups.

K = number of groups.

N = $n_1, n_2,$

Example-1

Suppose 3-teaching methods are to be tested on 9-students. The students (the experimental units) are randomly selected to the teaching methods (the treatment),3-students for each method. At the end of the training period students take a standardized test, and achieved the following scores:

1	2	3
109	110	112
108	115	107
104	108	108

Q: The question is to construct the ANOVA for the test.

Solution in steps:

1. We will arrange the required values in the table as shown:

Group-1 $n_1=3$		Group-2 $n_2=3$		Group-3 $n_{3=3}$	
X_1	X_1^2	X_2	X_2^2	X_3	X_3^2
109	11881	110	12100	112	12544
108	11664	115	13225	107	11449
104	10816	108	11664	108	11664
$\Sigma X_1 =321$	$\Sigma X_1^2=34361$	$\Sigma X_2=333$	$\Sigma X_2^2=36989$	$\Sigma X_3=327$	$\Sigma X_3^2-=35657$

2. We need to calculate the following values:

$N = n_1 + n_2 + n_3 = 9$.

$\Sigma X_T = \Sigma X_1 + \Sigma X_2 + \Sigma X_3$
    $= 321 + 333 + 327 = 981$

$\Sigma X_T^2 = \Sigma X_1^2 + \Sigma X_2^2 + \Sigma X_3^2$
    $= 34361 + 36989 + 35657 = 107007$

K = number of groups = 3

SS_B = sample squares between the groups.

SS_B = Sum of all groups $\{(\Sigma X_i)^2 / n_i) - (\Sigma X_T)\}/ N$

Sum of all groups $\{(\Sigma X_i)^2 / n_i\} = (\Sigma X_1)^2 / n_1 + (\Sigma X_2)^2 / n_2 + (\Sigma X_3)^2 / n_3$
    $= (321)^2 / 3 + (333)^2 / 3 + (327)^2 / 3$
    $= 34347 + 36963 + 35647 = 106953$
$(\Sigma X_T)^2 / N) = (981)^2 / 9 = 106929$

Then, $SS_B = 106953 - 106929 = 24$

SS_T = total sum of squares $= \Sigma X_T^2 - (\Sigma X_T)^2/N$
    $= 107007 - (981)^2 / 9 = 78$
SS_W = sample square within the groups

$SS_W = SS_T - SS_B = 78 - 24 = 54$

The ratio test $= F = \dfrac{MS_B}{MS_W}$

$MS_B = SS_B / df_B = SS_B / k\text{-}1 = 24/2 = 12$

$MS_W = SS_W / df_W = SS_W / N - k = 54 / 9\text{-}3 = 54 / 6 = 9$ →Then $F = 12/9 = 1.33$

The ANOVA table (general template) is:

Source	df	Sum of squares	Mean squares
Between groups	k-1	SS_B	MS_B
Within groups	N-K	SS_W	MS_W
Total			

Where, df=degree of sum

SS_b = sum of squares between the groups.

SS_w= Sum of squares within the groups.

K = number of groups. N = n_1, n_2, n_3.

Based on this model, we can construct the table for our problem as:

Source	df	Sum of squares	Mean squares
Between groups	2	24	12
Within groups	6	54	9
Total	**8**	**78**	**21**

To check our work using TI-83+ or TI-84+:

1. STAT → EDIT → enter data in L1, L2, L3 for the 3-groups
2. STAT → TESTS → F: ANOVA (l1, l2, l3)

Calculator will display the following:

One-Way-ANOVA
F=1.33
P=.3318
Factor
df=2
SS=24
MS=12
Error
df=6
SS= 54
MS=9
S_{xp} =3

When testing the difference among three or more means using F-test, the technique is called "Analysis of Variance".

Q: Suppose we want to test the claim that there is no difference in the 3-groups in example-1 above, with α = 0.05.

| Solution in steps: |

Using the table we can compute the sample mean of the 3-groups as shown below:

1	2	3
109	110	112
108	115	107
104	108	108
$\overline{X_1} = 107$	$\overline{X_2} = 111$	$\overline{X_3} = 109$
$S_1^2 = 7.02$	$S_2^2 = 12.96$	$S_3^2 = 7.02$

1. State the hypothesis for the 3-means:
   $H_0 : \mu_1 = \mu_2 = \mu_3$ (claim)
   $H_a :$ At least one mean is different.
2. Find the critical value (CV) using k=3, and N=9, and finding the 3-degrees of freedom first:
   df-of numerator = k-1 = 3-1 = 2
   df-of denominator = N-k = 9 – 3 = 6.
   Then with α=0.05 and using table for F-distribution gives:

	α = 0.05	
	df- N	
f-D	1	2
		↓
6 ——→		4.26

3. Compute the test value (TV) or the F-test value using the formula: $F = \dfrac{S^2\text{-between}}{S^2\text{-Within}}$

To get F we need to find S_B^2, and S_W^2 as follows:
Over all mean is = (109+108+104+110+115+108+112+107+108)/9 = 109.
Between group variance = S^2 = 3(107-

107	$(107-109)^2 = 4$	3(4) = 12
111	$(11-109)^2 = 4$	3(4)=12
109	$(109-109)^2 = 0$	0
		$\sum = 24$

Variance between groups is: $S_B{}^2 = 24/2 = 12$ (variance between)

Variance within groups is:

$S_W{}^2 = (3-1)(7.02) + (3-1)(12.96) + (3-1)(7.02) / (3-1) +(3-1)+(3-1)= 54 / 4 = 13.5$

Then F = 12/13.5 =0.9

4. Make the decision: Since $0.8 < 4.26 \rightarrow$ Do not reject the null hypothesis

9.2 Practice Problem

Suppose 4-teaching methods are to be tested on 20-students. The students (the experimental units) are randomly selected to the teaching methods (the treatment) 5-students for each method. At the end of the training period students take a standardized test, and achieved the following scores:

1	2	3	4
105	110	112	111
103	115	107	112
109	105	110	109
106	107	106	110
104	103	108	103

Construct the ANOVA for the test.

Index

R

S

T

U

V

W

Z

Table- 1: The Standard Normal Distribution
Z-Scores / Cumulative Area to the left of the Normal Curve

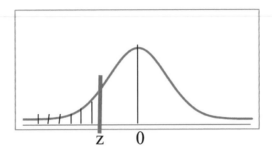

Z	0	0.01	0.02	0.03	0.04	0.05	0.06	0.07	0.08	0.09
0	0.5	0.504	0.508	0.512	0.516	0.5199	0.5239	0.5279	0.5319	0.5359
0.1	0.5398	0.5438	0.5478	0.5517	0.5557	0.5596	0.5636	0.5675	0.5714	0.5753
0.2	0.5793	0.5832	0.5871	0.591	0.5948	0.5987	0.6026	0.6064	0.6103	0.6141
0.3	0.6179	0.6217	0.6255	0.6293	0.6331	0.6368	0.6406	0.6443	0.648	0.6517
0.4	0.6554	0.6591	0.6628	0.6664	0.67	0.6736	0.6772	0.6808	0.6844	0.6879
0.5	0.6915	0.695	0.6985	0.7019	0.7054	0.7088	0.7123	0.7157	0.719	0.7224
0.6	0.7257	0.7291	0.7324	0.7357	0.7389	0.7422	0.7454	0.7486	0.7517	0.7549
0.7	0.758	0.7611	0.7642	0.7673	0.7704	0.7734	0.7764	0.7794	0.7823	0.7852
0.8	0.7881	0.791	0.7939	0.7967	0.7995	0.8023	0.8051	0.8078	0.8106	0.8133
0.9	0.8159	0.8186	0.8212	0.8238	0.8264	0.8289	0.8315	0.834	0.8365	0.8389
1	0.8413	0.8438	0.8461	0.8485	0.8508	0.8531	0.8554	0.8577	0.8599	0.8621
1.1	0.8643	0.8665	0.8686	0.8708	0.9729	0.8749	0.877	0.879	0.881	0.883
1.2	0.8849	0.8869	0.8888	0.8907	0.8925	0.8944	0.8962	0.898	0.8997	0.9015
1.3	0.9032	0.9049	0.9066	0.9082	0.9099	0.9115	0.9131	0.9147	0.9162	0.9177
1.4	0.9192	0.9207	0.9222	0.9236	0.9251	0.9265	0.9279	0.9292	0.9306	0.9319
1.5	0.9332	0.9345	0.9357	0.937	0.9382	0.9394	0.9404	0.9418	0.9429	0.9441
1.6	0.9452	0.9463	0.9474	0.9484	0.9495	0.9505	0.9515	0.9525	0.9535	0.9545
1.7	0.9554	0.9564	0.9573	0.9582	0.9591	0.9599	0.9608	0.9616	0.9625	0.9633
1.8	0.9641	0.9649	0.9656	0.9664	0.9671	0.9678	0.9686	0.9693	0.9699	0.9706
1.9	0.9713	0.9719	0.9726	0.9732	0.9738	0.9744	0.975	0.9756	0.9761	0.9767
2	0.9772	0.9778	0.9783	0.9788	0.9793	0.9798	0.9803	0.9808	0.9812	0.9817
2.1	0.9821	0.9826	0.983	..9834	0.9838	0.9842	0.9846	0.985	0.9854	0.9857
2.2	0.9861	0.9864	0.9868	0.9871	0.9875	0.9878	0.9881	0.9884	0.9887	0.989
2.3	0.9893	0.9896	0.9898	0.9901	0.9904	0.9906	0.9909	0.9911	0.9913	0.9916
2.4	0.9918	0.992	0.9922	0.9925	0.9927	0.9929	0.9931	0.9932	0.9934	0.9936

2.5	0.9938	0.994	0.9941	0.9943	0.9945	0.9946	0.9948	0.9949	0.9951	0.9952
2.6	0.9953	0.9955	0.9956	0.9957	0.9959	0.996	0.9961	0.9962	0.9963	0.9964
2.7	0.9965	0.9966	0.9967	0.9968	0.9969	0.997	0.9971	0.9972	0.9973	0.9974
2.8	0.9974	0.9975	0.9976	0.9977	0.9977	0.9978	0.9979	0.9979	0.998	0.9981
2.9	0.9981	0.9982	0.9982	0.9983	0.9984	0.9984	0.9985	0.9985	0.9986	0.9986
3	0.9987	0.9987	0.9987	0.9988	0.9988	0.9989	0.9989	0.9989	0.999	0.999
3.1	0.999	0.9991	0.9991	0.9991	0.9992	0.9992	0.9992	0.9992	0.9993	0.9993
3.2	0.9993	0.9993	0.9994	0.9994	0.9994	0.9994	0.9994	0.9995	0.9995	0.9995
3.3	0.9995	0.9995	0.9995	0.9996	0.9996	0.9996	0.9996	0.9996	0.9996	0.9997
3.4	0.9997	0.9997	0.9997	0.9997	0.9997	0.9997	0.9997	0.9997	0.9997	0.9998

<u>Table-2</u>: T-test Critical Values
        d.f. = degree of freedom = n-1
        1T = one tail test
        2T = two tail test

	CI	80%	90%	95%	98%	99%
	1T, α	0.1	0.05	0.025	0.01	0.005
d.f.	2T, α	0.2	0.1	0.05	0.02	0.01
1		3.078	6.314	12.706	31.821	63.657
2		1.886	2.92	4.303	6.965	9.925
3		1.638	2.353	3.182	4.541	5.841
4		1.533	2.132	2.776	3.747	4.604
5		1.476	2.015	2.571	3.365	4.032
6		1.44	1.943	2.447	3.143	3.707
7		1.415	1.895	2.365	2.998	3.499
8		1.397	1.86	2.306	2.896	3.355
9		1.383	1.833	2.262	2.821	3.25
10		1.372	1.812	2.228	2.764	3.169
11		1.363	1.796	2.201	2.718	3.106
12		1.356	1.782	2.179	2.681	3.055
13		1.35	1.771	2.16	2.65	3.012
14		1.345	1.761	2.145	2.624	2.977
15		1.341	1.753	2.131	2.602	2.947
16		1.337	1.746	2.12	2.583	2.921
17		1.333	1.74	2.11	2.567	2.898
18		1.33	1.734	2.101	2.552	2.878
19		1.328	1.729	2.093	2.539	2.861
20		1.325	1.725	2.086	2.528	2.845
21		1.323	1.721	2.08	2.518	2.831
22		1.321	1.717	2.074	2.508	2.819
23		1.319	1.714	2.069	2.5	2.807
24		1.318	1.711	2.064	2.492	2.797
25		1.316	1.708	2.06	2.485	2.787
26		1.315	1.706	2.056	2.479	2.779
27		1.314	1.703	2.052	2.473	2.771
28		1.313	1.701	2.048	2.467	2.763

29	1.311	1.699	2.045	2.462	2.756
30	1.31	1.697	2.042	2.457	2.75
32	1.309	1.694	2.037	2.449	2.738
34	1.307	1.691	2.032	2.441	2.728
36	1.306	1.688	2.028	2.434	2.719
38	1.304	1.686	2.024	2.429	2.712
40	1.303	1.684	2.021	2.423	2.704
45	1.301	1.679	2.014	2.412	2.69
50	1.299	1.676	2.009	2.403	2.678
55	1.297	1.673	2.004	2.396	2.668
60	1.296	1.671	2	2.39	2.66
65	1.295	1.669	1.997	2.385	2.654
70	1.294	1.667	1.994	2.381	2.648
75	1.293	1.665	1.992	2.377	2.643
80	1.292	1.664	1.99	2.374	2.639
90	1.291	1.662	1.987	2.368	2.632
100	1.29	1.66	1.984	2.364	2.626
500	1.283	1.648	1.965	2.334	2.586
1000	1.282	1.646	1.962	2.33	2.581
∞	1.82	1.65	1.96	2.35	2.58

Table – 3: χ^2- Distribution (area to the right of the curve)
Curve Skewed to the right

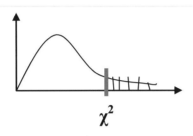

χ^2

d.f.	0.995	0.99	0.975	0.95	0.99	0.1	0.05	0.025	0.005
1			0.001	0.004	0.016	2.706	3.841	5.024	6.635
2	0.01	0.02	0.051	0.103	0.211	4.605	5.991	7.378	9.21
3	0.072	0.115	0.216	0.352	0.584	6.251	7.815	9.348	11.345
4	0.207	0.297	0.484	0.711	1.064	7.779	9.488	11.143	13.277
5	0.412	0.554	0.831	1.145	1.61	9.236	11.071	12.833	15.086
6	0.676	0.872	1.237	1.635	2.204	10.645	12.592	14.449	16.812
7	0.989	1.239	1.69	2.167	2.833	12.017	14.067	16.013	18.475
8	1.344	1.646	2.18	2.733	3.49	13.362	15.507	17.535	20.09
9	1.735	2.088	2.7	3.325	4.168	14.684	16.919	19.023	21.666
10	2.156	2.558	3.247	3.94	4.865	15.987	18.307	20.483	23.209
11	2.603	3.053	3.816	4.578	5.578	17.275	19.675	21.92	24.725
12	3.074	3.571	4.404	5.226	6.304	18.549	21.026	23.337	26.217
13	3.565	4.107	5.009	5.892	7.042	19.812	22.362	24.736	27.688
14	4.075	4.66	5.629	6.571	7.79	21.064	23.685	26.119	29.141
15	4.601	5.229	6.262	7.261	8,547	22.307	24.996	27.488	30.578
16	5.142	5.812	6.908	7.962	9.312	23.542	26.296	28.845	32
17	5.697	6.408	7.564	8.672	10.085	24.587	27.587	30.191	33.409
18	6.265	7.015	8.231	9.39	19.865	25.989	28.869	31.526	34.805
19	6.844	7.633	8.907	10.117	11.651	27.204	30.144	32.852	36.191
20	7.434	8.26	9.591	10.851	12.443	28.412	31.41	34.17	37.566
21	8.034	8.897	10.283	11.591	13.24	29.615	32.671	35.479	38.932
22	8.643	9.542	10.982	12.338	14.042	30.813	33.924	36.781	40.289
23	9.262	10.196	11.689	13.091	14.848	32.007	35.172	38.076	41.638
24	9.886	10.856	12.401	13.848	15.659	33.196	36.415	39.364	42.98
25	10.52	11.524	13.12	14.611	16.473	34.382	37.652	40.646	44.314

The header "α" spans across the table above the d.f. columns.

26	11.16	12.198	13.844	15.379	17.292	35.563	38.885	41.923	45.642
27	11.808	12.879	14.573	16.151	18.114	36.741	40.113	43.194	46.963
28	12.461	13.565	15.308	16.928	18.939	37.916	41.337	44.461	48.278
29	13.121	14.257	16.047	17.708	19.768	39.087	42.557	45.722	49.588
30	13.787	14.954	16.791	18.493	20.599	40.256	43.773	46.979	50.892
40	20.707	22.164	24.433	26.509	29.051	51.805	55.758	59.342	63.691
50	27.991	29.707	32.357	34.764	37.689	63.167	67.505	71.42	76.154
60	35.534	37.485	40.482	43.188	46.459	74.397	79.082	83.298	88.379
70	43.275	45.442	48.758	51.739	55.329	85.527	90.531	95.023	100.425
80	51.172	53.54	57.153	60.391	64.278	96.578	101.879	106.629	112.329
90	59.196	61.754	65.647	69.126	73.291	107.565	113.145	118.136	124.116
100	67.328	70.065	74.222	77.929	82.358	118.498	124.342	129.561	135.807

Table – 4: This table shows the Hypothesis –testing and Critical-Values:

Left-Tail	Two-Tail	Right-Tail
$H_0: \mu = k$ $H_a: \mu < k$	$H_0: \mu = k$ $H_a: \mu \neq k$	$H_0: \mu = k$ $H_a: \mu > k$
$\alpha = 0.10 \rightarrow C.V = -1.28$ $\alpha = 0.05 \rightarrow C.V = -1.65$ $\alpha = 0.01 \rightarrow C.V = -2.33$	$\alpha = 0.10 \rightarrow C.V = \pm 1.65$ $\alpha = 0.05 \rightarrow C.V = \pm 1.96$ $\alpha = 0.01 \rightarrow C.V = \pm 2.58$	$\alpha = 0.10 \rightarrow C.V = +1.28$ $\alpha = 0.05 \rightarrow C.V = +1.65$ $\alpha = 0.01 \rightarrow C.V = +2.33$

Formulas

Descriptive Statistics

Mean for a sample: $\bar{X} = \dfrac{X_1 + X_2 + X_3 + \ldots}{n} = \dfrac{\sum X}{n}$

Mean for population: $\mu = \dfrac{X_1 + X_2 + X_3 + \ldots}{N} = \dfrac{\sum X}{N}$

Mean for grouped frequency $\bar{X} = \dfrac{\sum f \cdot X_m}{n}$

Sample Variance $S^2 = \dfrac{\sum (\bar{X} - X)^2}{n-1}$.

Short –Cut- Formula $S^2 = \dfrac{n[\sum f \cdot X^2{}_m] - (\sum f \cdot X_m)^2}{n(n-1)}$
For sample variance

Probability

$P(E) = \dfrac{n(E)}{n(s)} = \dfrac{number\ of\ outcomes\ in\ the\ event\ E}{number\ of\ oucomes\ in\ the\ sample\ s}$

1.Rules of Addition:

- $P(A \cup B) = P(A) + P(B)$ If A and B are mutually exclusive
- $P(A \cup B) = P(A) + P(B) - P(A \cap B)$ If A and B are inclusive

2.Rules of Multiplication:

- $P(A \cap B) = P(A) \cdot P(B)$ If A, and B are independent.
- $P(A \cap B) = P(A) \cdot P(B/A)$ if A, and B are dependent

$$\text{Permutation} = nPr = p(n,r) = \frac{n!}{(n-r)!} = \frac{n(n-1)(n-2)\ldots 1}{(n-r)!}$$

$$\text{Combination} = nCr = C(n,r) = \frac{n!}{(n-r)!r!} = \frac{n(n-1)(n-2)\ldots 1}{(n-r)!\, r!}$$

$$\text{Empirical Probability } P(E) = \frac{Frequency\ of\ the\ class}{Total\ frequancy\ of\ the\ distribution} = \frac{f}{n}$$

Probability Distribution

$$\text{Mean} = \mu = \Sigma\, X\, P(X) = E(X)$$

$$\text{Variance} = \sigma^2 = \Sigma\,(X - \mu)^2 - \mu^2$$

$$\text{Standard Deviation} = \sigma = \sqrt{\Sigma\,(X - \mu)^2 - \mu^2}$$

$$\text{Binomial Probability } P(X) = \frac{n!}{(n-X)!\, X!}\, P^x\, q^{n-x}$$

Where, X= number of success in each trial, $0 \le X \le n$.

$$\text{Mean} = \mu = n\,P$$

$$\text{Variance} = \sigma^2 = n\,p\,q$$

$$\text{Standard Deviation} = \sigma = \sqrt{npq}$$

Normal Distribution

$$P(X < a) = P\left(Z < \frac{X - \mu}{\sigma}\right) = P(Z < b) \qquad \ldots (5)$$

$$Z = \frac{X - mean\ of\ pop}{Standard\ deviation\ of\ pop} = \frac{X - \mu}{\sigma}$$

Pearson Coefficient PC = $\dfrac{3(\bar{X} - \text{Median})}{S}$

Confidence Intervals

Confidence Intervals for the Mean(known variance):

CI is → $\bar{X} - Z_{\alpha/2}\,(\sigma/\sqrt{n}) \;<\; \mu \;<\; \bar{X} + Z_{\alpha/2}\,(\sigma/\sqrt{n})$

Confidence Intervals for the Mean(unknown variance):

$\bar{X} - t_{\alpha/2}\,(s/\sqrt{n}) \;<\; \mu \;<\; \bar{X} + t_{\alpha/2}\,(s/\sqrt{n})$, with df = n-1

confidence interval for the standard deviation:

CI → $\sqrt{\dfrac{S^2(n-1)}{\chi^2_R}} \;<\; \sigma \;<\; \sqrt{\dfrac{S^2(n-1)}{\chi^2_L}}$

Hypothesis Testing

z-test when σ is known $z = \dfrac{x - \mu}{(\sigma/\sqrt{n})}$

t-test when σ is unknown $t = \dfrac{\bar{x} - \mu}{(s/\sqrt{n})}$

n≥30, or n<30 if pop is normally distributed

Testing for Chi-Square:

z-test when np, nq ≥ 5 $z = \dfrac{\bar{P} - p}{\sqrt{pq}/\sqrt{n}}$

χ^2-test : $\chi^2 = \dfrac{(n-1)\,s^2}{\sigma^2}$

$$\chi^2_L = \frac{(n-1)\,S^2}{\chi^2\,(\,n-1,\,\alpha/2)} \quad , \quad \chi^2_R = \frac{(n-1)\,S^2}{\chi^2\,(\,n-1,\,1-\alpha/2)}$$

Testing between two means:

$$Z = \frac{(\bar{X}_1 - \bar{X}_2) - (\mu_1 - \mu_2)}{\sqrt{\dfrac{\sigma_1^2}{n_1} - \dfrac{\sigma_2^2}{n_2}}}$$

Linear correlation and Regression

Formula for r :

$$r = \frac{n\,(\Sigma xy) - (\Sigma x)\,(\Sigma y)}{\sqrt{[\,n(\Sigma x^2) - (\Sigma x)^2\,]\,[\,n(\Sigma y^2) - (\Sigma y)^2\,]}} \quad \ldots (1)$$

T-test formula:

$$t = r\,\sqrt{\frac{n-2}{1-r^2}} \quad \ldots (2)$$